Kousik Das Malakar

Demography of New Zealand

Kousik Das Malakar

Demography of New Zealand

Demographic Scenario, Social demography, and Islands demography

JustFiction Edition

Imprint
Any brand names and product names mentioned in this book are subject to trademark, brand or patent protection and are trademarks or registered trademarks of their respective holders. The use of brand names, product names, common names, trade names, product descriptions etc. even without a particular marking in this work is in no way to be construed to mean that such names may be regarded as unrestricted in respect of trademark and brand protection legislation and could thus be used by anyone.

Cover image: www.ingimage.com

Publisher:
JustFiction! Edition
is a trademark of
International Book Market Service Ltd., member of OmniScriptum Publishing Group
17 Meldrum Street, Beau Bassin 71504, Mauritius
Printed at: see last page
ISBN: 978-620-0-49470-2

Demography of New Zealand

Demographic Scenario, Social demography, Islands demography

Kousik Das Malakar

Jawaharlal Nehru University (JNU), New Delhi, India

Just Fiction! Edition

Founded in Germany

Some other books by the same Author

- Environmental impact of the COVID-19 pandemic: A lesson for the future.
- The Upper-Caste Community of Malakar and their hereditary livelihoods.
- Scope and Contents of Digital Humanities.
- Population and Environment: A Review (Part-6).
- Population and Environment: A Review (Part-5).
- Population and Environment: A Review (Part-4).
- Population and Environment: A Review (Part-3).
- Population and Environment: A Review (Part-2).
- Population and Environment: A Review (Part-1).
- A Regional Study of Bankura District in West Bengal, India.
- Digital humanities of India.
- Physical Characteristics of Baragere Village, Kharagpur-2, W.B.
- Socio-Economic Conditions of Baragere Village, Medinipur West, West Bengal.
- Agricultural Conditions of Baragere Village, Medinipur West, West Bengal.
- The Mangrove: Future of the Global seacoast.
- Project Field Report: Rural Development of India (Physical and Socio-Economic Conditions of Baragere Village, Kharagpur-II, Paschim Medinipur, West Bengal).
- A Basic Outline of Geomorphic Studies (Part-2).
- A Basic Outline of Geomorphic Studies (Part-1).
- Coronavirus (COVID-19): A Pandemic Situation in the World.
- Geospatial Technology: An Overview Concept of RS, GIS, and GPS.
- A Basic Outline of Population Studies.
- Climate Change: A Thinkable Manifesto of Future.
- Climatic Data Analysis and Regional Specialty (Part - II).
- Climatic Data Analysis and Regional Specialty (Part - I).
- Regional Identity of India (Part - 2).
- Regional Identity of India (Part – I).
- Agricultural Infrastructure and Land Use in India.
- River Basin Hydrological Measurement Technique.
- Social Studies: A Study of Public Health in India.
- Regional Studies: Some thinkable subjective discussion.

Dedications

My dear Parents

PREFACE

The book *"Demography of New Zealand"* is comprehensive and easy to understand a text. This book began with the concept of Introduction to New Zealand, Basic Concepts of demography, demographics Methods, World demographic Profile, demographic Profiles of the Populated Countries in the World, demography of New Zealand, Social demography of New Zealand, demographics of Auckland and the Cook Islands, Statistics New Zealand, and 2018 New Zealand census.

At least, I humbly hope that this study will contribute to increasing the concepts of demography of New Zealand. Suggestions for improving the book are always welcome and will be incorporated in the next edition.

JNU, New Delhi.   **Kousik Das Malakar**
September 04th, 2020

ACKNOWLEDGEMENT

As I was finalizing the manuscript of this book, my memory goes back to identify all those Physical and human geographers who positively affected me throughout my career and in many cases, they were not aware of the impact they were having on me. They range from my graduate Professors at Jawaharlal Nehru University (JNU) and Vidyasagar University (VU) in my early days to present geographers, Professors, colleagues, friends all over the country. To all of them, I express my regard and gratitude.
In writing this book, I have got immense help from a vast range of publications, So, I thank all the authors of the publications.
And I also grateful to my parents, because they are good wishes to me.

JNU, New Delhi
September 04th, 2020

Kousik Das Malakar

CONTENTS

Chapter – 1

Introduction to New Zealand

New Zealand is an island country in the south-western Pacific Ocean. It comprises two main landmasses the North Island and the South Island and around 600 smaller islands, covering a total area of 268,021 square kilometres. New Zealand is about 2,000 kilometres east of Australia across the Tasman Sea and 1,000 kilometres south of the islands of New Caledonia, Fiji, and Tonga. The country's varied topography and sharp mountain peaks, including the Southern Alps, owe much to tectonic uplift and volcanic eruptions. New Zealand's capital city is Wellington, and its most populous city is Auckland.

A developed country, New Zealand ranks highly in international comparisons of national performance, such as quality of life, education, protection of civil liberties, government transparency, and economic freedom. New Zealand underwent major economic changes during the 1980s, which transformed it from a protectionist to a liberalised free-trade economy. The service sector dominates the national economy, followed by the industrial sector, and agriculture; international tourism is a significant source of revenue. Nationally, legislative authority is vested in an elected, unicameral Parliament, while executive political power is exercised by the Cabinet, led by the Prime Minister, currently Jacinda Ardern. Queen Elizabeth II is the country's monarch and is represented by a governor-general, currently Dame Patsy Reddy. In addition, New Zealand is organised into 11 regional councils and 67 territorial authorities for local government purposes. The Realm of New Zealand also includes Tokelau (a dependent territory); the Cook Islands and Niue (self-governing states in free association with New Zealand); and the Ross Dependency, which is New Zealand's territorial claim in Antarctica.

Terminology:

The name derives from the 'kiwi', a native flightless bird, which is the national symbol of New Zealand. The Māori loanword 'Pakeha' usually refers to New Zealanders of European descent, although some reject this appellation, and some Maori uses it to refer to all non-Polynesian New Zealanders.

Location:

New Zealand (fig. 1) is an island country in the southwestern Pacific Ocean. It comprises two main landmasses—the North Island and the South Island and around 600 smaller islands, covering a total area of 268,021 square kilometres. New Zealand is about 2,000 kilometres east of Australia across the Tasman Sea and 1,000 kilometres south of the islands of New Caledonia, Fiji, and Tonga. The country's varied topography and sharp mountain peaks, including the Southern Alps, owe much to tectonic uplift and volcanic eruptions. New Zealand's capital city is Wellington, and its most populous city is Auckland.

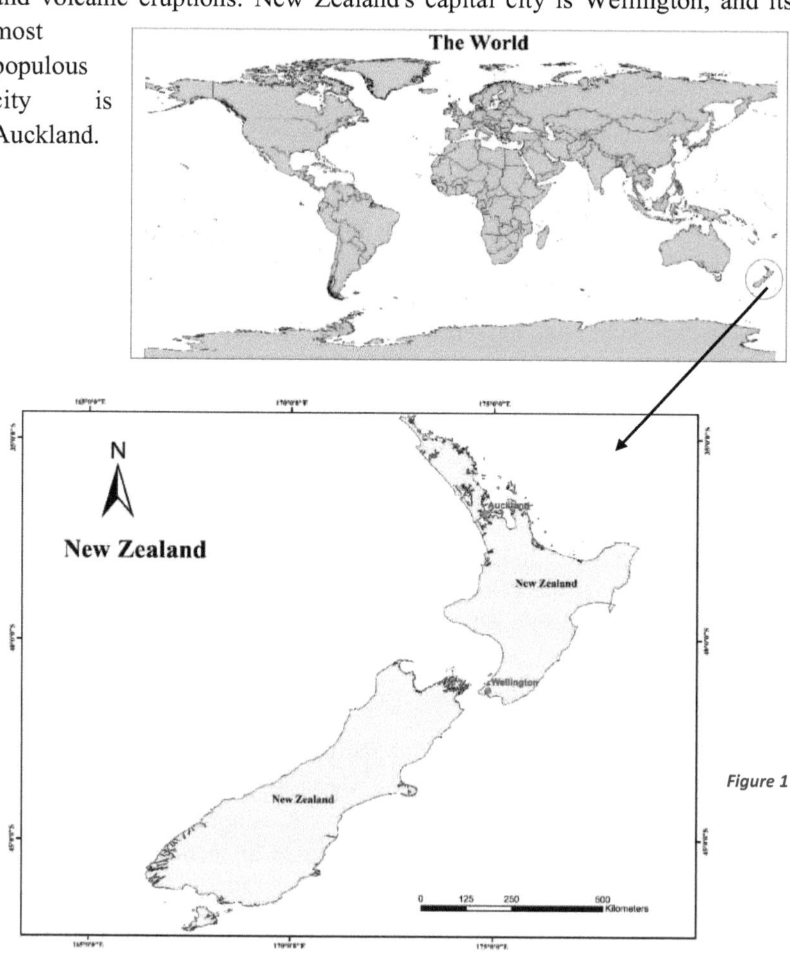

Figure 1

Geography and environment:

New Zealand is located near the centre of the water hemisphere and is made up of two main islands and a number of smaller islands. The two main islands (the North Island, or Te Ika-a-Māui, and the South Island, or Te Waipounamu) are separated by Cook Strait, 22 kilometres wide at its narrowest point. Besides the North and South Islands, the five largest inhabited islands are Stewart Island (across the Foveaux Strait), Chatham Island, Great Barrier Island (in the Hauraki Gulf), D'Urville Island (in the Marlborough Sounds) and Waiheke Island (about 22 km from central Auckland).

Table 1. Average daily maximum and minimum temperatures for the six largest cities of New Zealand

Location	Jan/Feb (°C)	Jan/Feb (°F)	July (°C)	July (°F)
Auckland	23/16	74/60	14/7	58/45
Wellington	20/13	68/56	11/6	52/42
Christchurch	22/12	72/53	10/0	51/33
Hamilton	24/13	75/56	14/4	57/39
Tauranga	24/15	75/59	14/6	58/42
Dunedin	19/11	66/53	10/3	50/37

New Zealand's climate is predominantly temperate maritime (Koppen: Cfb), with mean annual temperatures ranging from 10 °C in the south to 16 °C in the north. Historical maxima and minima are 42.4 °C in Rangiora, Canterbury and −25.6 °C in Ranfurly, Otago. Conditions vary sharply across regions from extremely wet on the West Coast of the South Island to almost semi-arid in Central Otago and the Mackenzie Basin of inland Canterbury, and subtropical in Northland. Of the seven largest cities, Christchurch is the driest, receiving on average only 640 millimetres of rain per year and Wellington the wettest, receiving almost twice that amount. Auckland, Wellington and Christchurch all receive a yearly average of more than 2,000 hours of sunshine. The southern and southwestern parts of the South Island have a cooler and cloudier climate, with around 1,400–1,600 hours; the northern and northeastern parts of the South Island are the sunniest areas of the country and receive about 2,400–2,500 hours. The general snow season is early June until early October, though cold snaps can occur outside this season. Snowfall is common in the eastern and southern parts of the South Island and mountain areas across the country.

The table 1 below lists climate normals for the warmest and coldest months in New Zealand's six largest cities. North Island cities are generally warmest in February. South Island cities are warmest in January.

Biodiversity

New Zealand's geographic isolation for 80 million years and island biogeography has influenced evolution of the country's species of animals, fungi and plants. Physical isolation has caused biological isolation, resulting in a dynamic evolutionary ecology with examples of very distinctive plants and animals as well as populations of widespread species. About 82% of New Zealand's indigenous vascular plants are endemic, covering 1,944 species across 65 genera. The number of fungi recorded from New Zealand, including lichen-forming species, is not known, nor is the proportion of those fungi which are endemic, but one estimate suggests there are about 2,300 species of lichen-forming fungi in New Zealand and 40% of these are endemic. The two main types of forest are those dominated by broadleaf trees with emergent podocarps, or by southern beech in cooler climates. The remaining vegetation types consist of grasslands, the majority of which are tussock.

Before the arrival of humans, an estimated 80% of the land was covered in forest, with only high alpine, wet, infertile and volcanic areas without trees. Massive deforestation occurred after humans arrived, with around half the forest cover lost to fire after Polynesian settlement. Much of the remaining forest fell after European settlement, being logged or cleared to make room for pastoral farming, leaving forest occupying only 23% of the land.

The forests were dominated by birds, and the lack of mammalian predators led to some like the kiwi, kakapo, weka and takahē evolving flightlessness. The arrival of humans, associated changes to habitat, and the introduction of rats, ferrets and other mammals led to the extinction of many bird species, including large birds like the moa and Haast's eagle.

Other indigenous animals are represented by reptiles (tuatara, skinks and geckos), frogs, spiders, insects and snails. Some, such as the tuatara, are so unique that they have been called living fossils. Three species of bats (one since extinct) were the only sign of native land mammals in New Zealand until the 2006 discovery of bones from a unique, mouse-sized land mammal at least 16 million years old. Marine mammals however are abundant, with almost half the world's cetaceans (whales, dolphins, and porpoises) and large numbers of fur seals reported in New Zealand waters. Many seabirds breed in New Zealand, a third of them unique to the country. More penguin species are found in New Zealand than in any other country.

Since human arrival, almost half of the country's vertebrate species have become extinct, including at least fifty-one birds, three frogs, three lizards, one freshwater fish, and one bat. Others are endangered or have had their range severely reduced. However, New Zealand conservationists have pioneered several methods to help threatened wildlife recover, including island sanctuaries, pest control, wildlife translocation, fostering and ecological restoration of islands and other protected areas.

New Zealand

Capital	Wellington	Largest city	Auckland
Official languages	English, Maori, NZ Sign Language	**Ethnic groups (2018)**	71.8% European; 16.5% Māori; 15.3% Asian; 9.0% Pacific peoples; 1.5% ME/LA/African; 1.2% Other
Religion (2018)	48.5% No religion; 37.0% Christianity; 2.6% Hinduism; 1.3% Islam; 1.1% Buddhism; 1.9% Others; 6.6% Undeclared	**Demonym(s)**	New Zealander Kiwi (informal)
Government **Unitary**	Monarch: Elizabeth II Governor-General: Patsy Reddy	**Legislature**	Parliament (House of

parliamentary constitutional monarchy	Prime Minister: Jacinda Ardern		Representatives)
Stages of independence from the United Kingdom	Responsible government: 7 May 1856 Dominion: 26 September 1907 Statute of Westminster adopted: 25 November 1947	**Area**	Total: 268,021 km2 Water (%): 1.6
Population	August 2020 estimate: 5,028,980 (120th). 2018 census: 4,699,755. Density: 18.3/km2 (47.4/sq mi) (167th).	**GDP (PPP) 2018 estimate**	Total: $199 billion Per capita: $40,266
GDP (nominal) 2018 estimate	Total: $206 billion Per capita: $41,616	**Gini (2019)**	Negative increase 33.9
HDI (2018)	Increase 0.921: very high.	**Currency**	New Zealand dollar ($) (NZD)
Time zone	UTC+12 Summer (DST) UTC+13	**Source:**	Statistics New Zealand, 2018

****

Chapter – 2

Basic Concepts of Demography

Demographic thoughts traced back to antiquity, and were present in many civilisations and cultures, like Ancient Greece, Ancient Rome, China and India. Demography is made up of two word Demos and Graphy. The term Demography refers to the overall study of population.

In ancient Greece, this can be found in the writings of Herodotus, Thucidides, Hippocrates, Epicurus, Protagoras, Polus, Plato and Aristotle. In Rome, writers and philosophers like Cicero, Seneca, Pliny the elder, Marcus Aurelius, Epictetus, Cato, and Columella also expressed important ideas on this ground.

In the middle ages, Christian thinkers devoted much time in refuting the Classical ideas on demography. Important contributors to the field were William of Conches, Bartholomew of Lucca, William of Auvergne, William of Pagula, and Muslim sociologists like Ibn Khaldun.

One of the earliest demographic studies in the modern period was Natural and Political Observations Made upon the Bills of Mortality (1662) by John Graunt, which contains a primitive form of life table. Among the study's findings were that one third of the children in London died before their sixteenth birthday. Mathematicians, such as Edmond Halley, developed the life table as the basis for life insurance mathematics. Richard Price was credited with the first textbook on life contingencies published in 1771, followed later by Augustus de Morgan, 'On the Application of Probabilities to Life Contingencies' (1838).

In 1755, Benjamin Franklin published his essay Observations Concerning the Increase of Mankind, Peopling of Countries, etc., projecting exponential growth in British colonies. His work influenced Thomas Robert Malthus, who, writing at the end of the 18th century, feared that, if unchecked, population growth would tend to outstrip growth in food production, leading to ever-increasing famine and poverty. Malthus is seen as the intellectual father of ideas of overpopulation and the limits to growth. Later, more sophisticated and realistic models were presented by Benjamin Gompertz and Verhulst.

In 1855, a Belgian scholar Achille Guillard defined demography as the natural and social history of human species or the mathematical knowledge of populations, of their general changes, and of their physical, civil, intellectual and moral condition.

The period 1860-1910 can be characterised as a period of transition wherein demography emerged from statistics as a separate field of interest. This period included a panoply of international 'great demographers' like Adolphe Quételet (1796–1874), William Farr (1807–1883), Louis-Adolphe Bertillon (1821–1883) and his son Jacques (1851–1922), Joseph Körösi (1844–1906), Anders Nicolas Kaier (1838–1919), Richard Böckh (1824–1907), Émile Durkheim (1858-1917), Wilhelm Lexis (1837–1914), and Luigi Bodio (1840–1920) contributed to the development of demography and to the toolkit of methods and techniques of demographic analysis.

Demographic analysis can cover whole societies or groups defined by criteria such as education, nationality, religion, and ethnicity. Educational institutions usually treat demography as a field of sociology, though there are a number of independent demography departments. Based on the demographic research of the earth, earth's population up to the year 2050 and 2100 can be estimated by demographers.

Formal demography limits its object of study to the measurement of population processes, while the broader field of social demography or population studies also analyses the relationships between economic, social, cultural, and biological processes influencing a population.

Methodological Subdivisions:

There are two types of data collection; direct and indirect with several different methods of each type.

Direct methods

Direct data comes from vital statistics registries that track all births and deaths as well as certain changes in legal status such as marriage, divorce, and migration. In developed countries with good registration systems (such as the United States and much of Europe), registry statistics are the best method for estimating the number of births and deaths.

A census is the other common direct method of collecting demographic data. A census is usually conducted by a national government and attempts to enumerate every person in a country. In contrast to vital statistics data, which are typically collected continuously and summarized on an annual basis, censuses typically occur only every 10 years or so, and thus are not usually the best source of data on births and deaths. Analyses are conducted after a census to estimate how much over or undercounting took place. These compare the sex ratios from the census data to those estimated from natural values and mortality data.

Censuses do more than just count people. They typically collect information about families or households in addition to individual characteristics such as age, sex, marital status, literacy/education, employment status, and occupation, and geographical location. They may also collect data on migration (or place of birth or of previous residence), language, religion, nationality (or ethnicity or race), and citizenship. In countries in which the vital registration system may be incomplete, the censuses are also used as a direct source of information about fertility and mortality; for example the censuses of the People's Republic of China gather information on births and deaths that occurred in the 18 months immediately preceding the census.

Indirect methods

Indirect methods of collecting data are required in countries and periods where full data are not available, such as is the case in much of the developing world, and most of historical demography. One of these techniques in contemporary demography is the sister method, where survey researchers ask women how many of their sisters have died or had children and at what age. With these surveys, researchers can then indirectly estimate birth or death rates for the entire population. Other indirect methods in contemporary demography include asking people about siblings, parents, and children. Other indirect methods are necessary in historical demography.

There are a variety of demographic methods for modelling population processes. They include models of mortality (including the life table, Gompertz models, hazards models, Cox proportional hazards models, multiple decrement life tables, Brass relational logits), fertility (Hernes model, Coale-Trussell models, parity progression ratios), marriage (Singulate Mean at Marriage, Page model), disability (Sullivan's method, multistate life tables), population projections (Lee-Carter model, the Leslie Matrix), and population momentum.

The United Kingdom has a series of four national birth cohort studies, the first three spaced apart by 12 years: the 1946 National Survey of Health and Development, the 1958 National Child Development Study, the 1970 British Cohort Study, and the Millennium Cohort Study, begun much more recently in 2000. These have followed the lives of samples of people (typically beginning with around 17,000 in each study) for many years, and are still continuing. As the samples have been drawn in a nationally representative way, inferences can be drawn from these studies about the differences between four distinct generations of British people in terms of their health, education, attitudes, childbearing and employment patterns.

****

Chapter – 3

Demographics Methods

- The crude birth rate, the annual number of live births per 1,000 people.
- The general fertility rate, the annual number of live births per 1,000 women of childbearing age (often taken to be from 15 to 49 years old, but sometimes from 15 to 44).
- The age-specific fertility rates, the annual number of live births per 1,000 women in particular age groups (usually age 15–19, 20-24 etc.)
- The crude death rate, the annual number of deaths per 1,000 people.
- The infant mortality rate, the annual number of deaths of children less than 1 year old per 1,000 live births.
- The expectation of life (or life expectancy), the number of years that an individual at a given age could expect to live at present mortality levels.
- The total fertility rate, the number of live births per woman completing her reproductive life, if her childbearing at each age reflected current age-specific fertility rates.
- The replacement level fertility, the average number of children women must have in order to replace the population for the next generation. For example, the replacement level fertility in the US is 2.11.
- The gross reproduction rate, the number of daughters who would be born to a woman completing her reproductive life at current age-specific fertility rates.
- The net reproduction ratio is the expected number of daughters, per newborn prospective mother, who may or may not survive to and through the ages of childbearing.
- A stable population, one that has had constant crude birth and death rates for such a long period of time that the percentage of people in every age class remains constant, or equivalently, the population pyramid has an unchanging structure.

⬇ A stationary population, one that is both stable and unchanging in size (the difference between crude birth rate and crude death rate is zero).

⬇ A stable population does not necessarily remain fixed in size. It can be expanding or shrinking.

Note that the crude death rate as defined above and applied to a whole population can give a misleading impression. For example, the number of deaths per 1,000 people can be higher for developed nations than in less-developed countries, despite standards of health being better in developed countries. This is because developed countries have proportionally older people, who are more likely to die in a given year, so that the overall mortality rate can be higher even if the mortality rate at any given age is lower. A more complete picture of mortality is given by a life table, which summarizes mortality separately at each age. A life table is necessary to give a good estimate of life expectancy.

Mathematical Expression / Methods / Equations

Sl_No	Name	Mathematical Expression / Methods / Equations	Remarks
1	**Population growth rate (G)**	$G = [\{(P_1 - P_2)/P_1\}*100]$	G- Population growth rate. P_1- Population of the base year. P_2- Population of the present year.
2	**YADR (Young Age Dependency Ratio)**	[(Total population in 0 to 14 years age/ Total population in 15 to 59 years age)*100].	
3	**OADR (Old Age Dependency**	[(Total population in 60+ years age/ Total population in 15 to 59 years age)*100].	

	Ratio)		
4	**Sex Ratio**	(Male populations/Female populations) × K (1000 or 100)	Used the value of 'K' is based on Developed and Developed countries number of populations
5	**Birth rate**	(Number of Births / Total Populations) × 1000	
6	**Death rate**	(Number of deaths / Total Populations) × 1000	
7	**Total fertility rate**	This is an overall summary measure of fertility and is obtained by summing the age-specific fertility rate for each age of the childbearing span. In other words, the total fertility rate is the number of children that a woman of the hypothetical cohort would bear during her lifetime if she were to bear children throughout her life at the Age-Specific Fertility Rates for a given year and if none of them dies before crossing the age of reproduction. It is constructed as follows; **TFR = $\sum f_x$ [$\sum$ (five-year age-specific birth rates for females aged 10 to 49) × 5]** Therefore, the **age Specific Marital Fertility Rate (ASMFR):** ASMFR is measured as number of births per year in a given age group to the total number of married women in that age group at mid-year.	

		It is constructed as follows; $$\text{ASMFR} = \frac{nBx}{nW\text{m}x}$$ Where, 'nBx' is the number of births in a year to the married women of ages x to x+n years in a given years and geographical region; '$_nW^mx$' is the number of women aged x to x+n years at mid-year in a given year and geographical area.	
8	**Natural change between Crude birth rate and Crude death rate**	(Crude birth rate - Crude death rate)	
9	**Net migration**	$$= \frac{I-o}{P} \times 1000$$ Where, ❖ Rate of In-migration (Im) = $\frac{I}{P} \times 1000$ ❖ Rate of Out-migration (Em) $= \frac{O}{P} \times 1000$	I- In-migration; O- Out-migration; P- Place of Origin or Place of destination.
10	**Mother's mean age at first birth**	$$M(t) = \frac{\sum_0^{amax}(a + 0.5)\, b\,(a,t)}{\sum_0^{amax} b\,(a,\ t)}$$	Where, M(t) = Average age at first birth at time t b(a,t) = the age-specific birth rate for birth order one at (single) age a and time t. a_{max}= the

		highest age at which first births are observed.
11	**Maternal mortality rate**	(Number of resident maternal deaths/Number of resident live births) x 100,000
12	**Infant mortality rate**	[(number of deaths to live born infants under one year of age) / (number of births)] × 1000 Based on life table estimation, **Observed data:** $_nN_x$ = mid-year population in age interval x to x +n $_nD_x$ = deaths between ages x and x +n during the year
13	**Life expectancy at birth**	**Steps for period life table construction:** 1. $_nm_x \approx nMx = (_nD_x / _nN_x)$. 2. $_na_x$: calculated from Coale and Demeny equations. 3. $_nq_x = \{(n\times_nm_x) / (1+ (n - _na_x) \times _nm_x)\}$ $\infty q_{85} = 1.00$ 4. $_nP_x = 1- _nq_x$ 5. $l_0 = 100,000$ $l_{x+n} = lx \cdot _nP_x$ 6. $_nd_x = lx - l_{x+n}$ 7. $_nL_x = n \cdot l_{x+n} + _na_x \cdot _nd_x$ 8. $T_x = \sum_{a-x}^{\infty} nL_a$ 9. $e^0_x = T_x / l_x$
14	**Population projections**	**Cohort-component method:** Cohort-component method is widely used method for the population projections. Following the description of this method; **For female:** $_5N^F_x (t)$ = number of women aged x to x+5 at time t. $_5L^F_x$ = Number of person-years lived by women from age x to x + 5.

$_5F_x$ = age-specific fertility rate in interval x to x + 5.

$_5N\ ^F_x (t+5) = _5N\ ^F_{x-5}{}^{(t)} \times (_5L\ ^F_x / _5L\ ^F_{x-5})$;

$\infty N^F_{85} (t+5) = (5N^F_{80} (t) + \infty N^F_{80} (t)) . (T^F_{85} / T^F_{80})$;

$_5B_x [t, t+5] = 5 \times _5F_x \times [\{_5N\ ^F_x (t) + _5N\ ^F_x (t+5)\} / 2]$

= births to women aged x to x + 5 between time f and time t + 5.

$B^F \quad [t, \quad t+5] \quad = \sum_{x=a}^{\beta-5} 5Bx\ [t, t+5]\ \times \frac{1}{1} + 1.05$

= number of females births between t and t + 5 (with SRB = 1.05).

$_5N_0^F(t+5) = B^F [t, t+5] \times 5\ L_0^F / 5l_0$

For Males:

$_5N\ ^M_x (t)$ = number of women aged x to x+5 at time t.

$_5L\ ^M_x$ = Number of person-years lived by women from age x to x + 5.

$_5N\ ^M_x (t+5) = _5N\ ^M_{x-5}{}^{(t)} \times (_5L\ ^M_x / _5L\ ^M_{x-5})$;

$\infty N^M_{85} (t+5) = (5N^M_{80} (t) + \infty N^M_{80} (t)) . (T^M_{85} / T^M_{80})$;

$B^M \quad [t, \quad t+5] \quad = \sum_{x=a}^{\beta-5} 5Bx\ [t, t+5]\ \times \frac{1}{1} + 1.05$

= number of the males births between t and t + 5.

$_5N_0^M(t+5) = B^M [t, t+5] \times 5\ L_0^M / 5l_0$

15	**Calculation of percentage (i e, Religious, languages and other calculation for percentages)**	= (Individual values / Total Values) × 100

| 16 | **Unemployment Rate (U)** | To calculate the unemployment rate, the number of unemployed people is divided by the number of people in the labour force, which consists of all employed and unemployed people. The ratio is expressed as a percentage.

$$U = \frac{\text{the number of unemployed peoples}}{\text{the number of people in the labour force}} \times 100$$ |
| 17 | **Others** | |

Whipple's index:

Whipple's Index is a demographic technique of measuring age heaping proposed by the American demographer G.C. Whipple. Whipple Index measure the magnitude of preference and avoidance of an ending digit. The Whipple Index is one of the most widely applied index to assess age heaping. Modified Whipple's Index was developed by Noumbissi (1992) and later modified by Spoorenberg (2007) to measure the degree of preferences or avoidances for all the terminal digits. It is also used in the graphical analysis.

The Whipple's Index is the proportion of the total population in age category 23 and 62 who report their ages in terminal digits '0' and '5' to one-fifth of the entire population in the specified age band, multiplied by 100. This expression the relative preference for digits '0' and '5'. It is computed as following equation 1. For the 10 years range;

$$\textbf{Whipple's Index} = \frac{\sum(N30+N40+\cdots..+N60)}{\frac{1}{10}\sum(N23+N24+N25+\cdots......+N62)} \times 100$$

Where, N_x = denotes the number of individuals at age x.

The age range chosen from 23 to 62 is basically arbitrary. Young ages and old ages are often excluded, assuming they are more prone to other types of misreporting errors then by preference for particular ending digits. Younger age groups (0 to 4 years) are understated, whereas the older ages (60 and above) may be overstated and have few persons in these age group due to high mortality. If there is no heaping at the reporting of age terminating with '0' and '5', the index will take a value of 100. If all reported ages ends with '0' and '5' (that is complete heaping), the index will assumed maximum value of 500. The summarized Whipple's Index scale for evaluating age data accuracy are represent in table 2.

Table 2. Scale of Whipple's Index		
Sl_No.	WI Values	Quality of the Data
1	Less than 105	Highly Accurate
2	105 to 110	Fairly Accurate
3	110 to 125	Approximate
4	125 to 175	Rough
5	More than 175	Very Rough

Myers' blended index:

Myer's blended index is calculated for the age above 10 years and shows the excess or deficit of people in ages ending in any of the 10 digits expressed as percentages. It is based on the assumption that the population is equally distributed among the different ages. The steps in the calculation of Myers' blended index are as follows:

Step - 1

Sum of populations ending in each digit over the whole range starting with the lower limit of the range (e.g., 10, 20, 30, 40,....; 11, 21, 31,....).

Step - 2

Ascertain sum excluding the first population combined in step 1 (e.g., 20, 30, 40,....; 21, 31, 41,....).

Step - 3

Weight the sums in steps 1 and 2 and add the results to obtain a blended population (e.g., weights 1 and 9 for 0 digit, weights 2 and 8 for 1, etc.).

Step - 4

Convert distribution in step 3 into percentages.

Step - 5

Take the deviation of each percentage in step 4 from 10.0, which is the expected value for each percentage.

Step - 6

A summary index of preference for all terminal digits is derived as one half of the sum of the deviations from 10 %, each without regard to signs.

Direct Standardizations Method

A challenge to the indirect method of standardization came in 1883 from within the Registrar General's Office). Ogle proposed the use of what is now known as the direct method of standardization. The method can be considered as the opposite of indirect standardization. The type of information required on the standard population for the indirect method is required for the index population in the direct method and vice versa. Thus the information needed for calculating a directly standardized ratio is:

- age-specific rates in the index population;
- the size of the standard population in each age group;
- the total number of deaths (or cases of disease) in the standard population.

The formula for the directly standardized ratio, usually termed the Comparative Mortality Figure (CMF) when deaths are being considered, is as follows:

$$\frac{\sum N_i\, r_i}{D}$$

Analogous to the formula for the SMR, the CMF can be expressed as a ratio of the expected deaths in the standard population on the basis of index rates to the total number of deaths in the standard population; in other words,

CMF = E/D where, E= $\sum N_i\, r_i$

Multiplying by the crude rate in the standard population gives the standardized rate as follows

$$\frac{R \sum Ni\, ri}{D} = \frac{\sum Ni\, ri}{N},$$

since D/R = N.

Indirect Standardization method:

The information required for use of the indirect method is as follows:

- age-specific rates in a standard population;
- The size of the index population in each age group; and
- the total number of deaths (or cases of disease) in the index population.

The formula for the indirectly standardized ratio is;

$$\frac{d}{\sum ni\, Ri}$$

Such ratios are widely known as Standardized Mortality Ratios (SMR) when deaths have been studied. Similar names are adopted for morbidity, such as Standardized Incidence Ratios for cancer incidence rates. An alternate way of considering an indirectly standardized ratio is as a ratio of the observed number of events to the number expected in the index population on the basis of standard rates; in other words,

SMR = d/e where, e = $\sum ni\, Ri$

One can then obtain the indirectly standardized rate by multiplying the ratio by the crude rate in the standard population:

$$\frac{d\, R}{\sum ni\, Ri}$$

Example of Age-standardization:

Formula:

$ASCDR^{SW} = \sum_{i=1}^{\infty} Mi^{SW}. \; C^s_i$ =Age-standardized crude death rate for Sweden.

$ASCDR^{K} = \sum_{i=1}^{\infty} MKi. \; C^s_i$ = Age-standardized crude death rate for Kazakhstan.

$C^s_i = \{(C^{sw}_i + C^K_i) / 2\}$ = Average age distribution.

General Fertility Rate (GFR):

This measure addresses the crude nature of the CBR by focusing on that section of the population at risk of having births – women aged 15-49.

It is constructed as follows (equation no, 1):

$$\textbf{GFR} = \frac{B}{W\ 15-49} \times \textbf{K} \ \ldots\ldots \ \text{Eq. 1}$$

Where, 'B' is the total number of live births during a given year and geographical area; 'W15-49' is the total number of women of child bearing age 15-44 at the mid-point of the year in a given geographical area, and 'K' is a constant and usually taken as 1,000.

It can easily be constructed from vital registration, census, or survey data and does not rely on exact ages (except at the beginning and end of age range). The General Fertility Rate (GFR) has specific uses in certain circumstances when one wants to know the total number of births for all women in the fertile ages. The main advantage of the GFR is that it includes the female population in their reproductive ages who are supposed to be exposed to the risk of giving birth. This measure generally used in population projection using component projection method.

General Marital Fertility Rate (GMFR):

General Marital Fertility Rate is the overall measure of fertility of married women. It is defined as the number of births per year per thousand mid-year married women within reproductive ages.

It is constructed as follows (equation no, 2):

$$\text{GMFR} = \frac{B}{Wm\ 15-49} \times \textbf{K} \ \ \ldots\ldots \ \text{Eq. 2}$$

Where, 'B' is the total number of live births during a given year and geographical area; 'Wm 15-44' is the total number of married women of child bearing age 15-49 at the mid-point of the year, and 'K' is a constant and usually taken as 1,000.

Age-Specific Fertility Rate (ASFR):

The age pattern of child bearing in any population is the best revealed by computing age specific fertility rates. The age-specific fertility rates is the number of births per year per women in a given age group in a given year and geographical region.

It is constructed as follows (equation no, 3):

$$ASFR = \frac{_nB_x}{_nW_x} \quad \text{....... Eq. 3}$$

Where, '$_nB_x$' is number of births to women of ages x to x+n years in a given year and area. '$_nW_x$' is number of women aged x and x+n years at mid-year in a given year and area, and 'n' is usually taken as 5 years.

Age Specific Marital Fertility Rate (ASMFR):

ASMFR is measured as number of births per year in a given age group to the total number of married women in that age group at mid-year.

It is constructed as follows (equation no, 4):

$$ASMFR = \frac{nBx}{nWmx} \quad \text{....... Eq. 4}$$

Where, 'nBx' is the number of births in a year to the married women of ages x to x+n years in a given years and geographical region; '$_nW^mx$' is the number of women aged x to x+n years at mid-year in a given year and geographical area.

Total Fertility Rate (TFR):

This is an overall summary measure of fertility and is obtained by summing the age-specific fertility rate for each age of the childbearing span. In other words, the total fertility rate is the number of children that a woman of the hypothetical cohort would bear during her lifetime if she were to bear children throughout her life at the Age-Specific Fertility Rates for a given year and if none of them dies before crossing the age of reproduction.

It is constructed as follows (equation no, 5):

$$TFR = \sum f_x \quad \text{....... Eq. 5}$$

Where, 'x' for single year ASFR.

Total Marital Fertility Rate (TMFR):

This is an over-all summary measure of marital fertility and is obtained by summing the age-specific marital fertility rate for each age of the childbearing span. In other words, total marital fertility rate is the number of children which a married woman of hypothetical cohort would bear during her life time if she were to bear children throughout her married life at the Age Specific Marital Fertility Rates for given year and if none of them dies before crossing the age of reproduction.

It is constructed as follows (equation no, 6):

$$\text{TMFR} = 5 \times \sum {}_5g_x \quad \text{....... Eq. 6}$$

For 'x' = 15, 20, 25.......40 (so, 'n' = 5)

Life table estimation

Introduction:

The life table is one of the most important devices used in demography. Life table is a mathematical sample which gives a view of death in a country and is the basis for measuring the average life expectancy in a society. It tells about the probability of a person dying at a certain age, or living upto a definite age. According to Bogue, *"The life table is a mathematical model that portrays mortality condition at a particular time among a population and provides a basis for measuring longevity. It is based on age specific mortality rates observed for a population for a particular year."*

Thus a life table is a mathematical device that shows the life span of persons up to a particular age or their probable date of death relates to a cohort of people born at the same time until they die. A life table can be constructed for a country and an area on the basis of sex, occupation, race, etc. There are two types of life table;

- **Generation / Cohort Life Table:** The Cohort or Generation Life Table "summarises the age-specific mortality experience of a given birth cohort (a group of persons all born at the same time) for its life and thus extends over many calendar years."

- **Period Life Table:** The "Period Life Table summarises the age-specific mortality conditions pertaining to a given or another short time period."

Basic assumptions of the life table:

- A hypothetical cohort of life table usually comprises of 1,000 or 10,000 or 1, 00,000 births.
- The deaths are equally distributed throughout the year.
- The cohort of people diminishes gradually by death only.
- The cohort of life tables is generally constructed separately for males and females.
- The cohort of persons dies at a fixed age which does not change.
- There is no change in death rates in overtime.
- The death rate is related to a pre-determined age-specific death rate.
- The cohort is closed to the in-migration and out-migration.

Importance of Life Table:

Life tables have been constructed by Graunt, Reed and Merrell, Keyfitz, Greville, and other demographers for estimating population trends regarding death rates, average expectation of life, migration rates, etc.

Uses of life tables:

- The life table is used to project future population on the basis of the present death rate.
- The method of constructing a life table can be followed to estimate the cause of specific death rates, male and female death rates, etc.
- It helps in determining the average expectation of life-based on age-specific death rates.
- Life tables can be used to compare population trends at national and international levels.
- The survival rates in a life table can be used to calculate the net migration rate on the basis of age distribution at 5 or 10-year interval.
- Instead of a single life table, multiple decrement life tables relating to the cause-specific death rate, male and female death rates, etc. can be constructed for analyzing socio-economic data in a country.

- Life tables are particularly used for formulating family planning programs relating to infant mortality, maternal deaths, health programs, etc. They can also be used for evaluating family planning programs.
- By constructing a life table based on the age of marriage, marriage patterns and changes in them can be estimated.

Now a days, life tables are used by life insurance companies in order to estimate the average life expectancy of persons, separately for males and females. They help in determining the amount of premium to be paid by a person falling in a specific age group.

An Example of Life table construction:

Observed data:

$_nN_x$ = mid-year population in age interval x to x +n

$_nD_x$ = deaths between ages x and x +n during the year

Steps for period life table construction:

$_nm_x \approx nMx = (_nD_x / {_nN_x})$.

$_na_x$: calculated from Coale and Demeny equations.

$_nq_x = \{(n\times_nm_x) / (1+ (n - {_na_x}) \times {_nm_x})\} \quad \infty q_{85} = 1.00$

$_nP_x = 1 - {_nq_x}$

$l_0 = 100{,}000 \quad l_{x+n} = lx \cdot {_nP_x}$

$_nd_x = lx - l_{x+n}$

$_nL_x = n \cdot l_{x+n} + {_na_x} \cdot {_nd_x}$

$T_x = \sum_{a=x}^{\infty} nLa$

$e^0_x = T_x / l_x$

Estimation of Population Projection

Introduction:

Population projections are estimates of the population for future dates. They are typically based on an estimated population consistent with the most recent decennial census and are produced using the cohort-component method. In contrast with intercensal estimates and censuses, which usually involve some sort of field data gathering, projections usually involve mathematical models based only on pre-existing data may be made by a governmental organization, or by those unaffiliated with a government.

Basically, it is based on knowledge of the past trends, and, for the future, on assumptions made for three components: fertility, mortality and migration.

Assumption

When the cohort component method is used as a projection tool, it assumes the components of demographic change, mortality, fertility, and migration, will remain constant throughout the projection period.

Cohort-component method:

Cohort-component method is widely used method for the population projections. Following the description of this method;

For female:

$_5N\,^F_X\,(t)$ = number of women aged x to x+5 at time t.

$_5L\,^F_X$ = Number of person-years lived by women from age x to x + 5.

$_5F_X$ = age-specific fertility rate in interval x to x + 5.

$_5N\,^F_X\,(t+5) = {}_5N\,^F_{X-5}{}^{(t)} \times ({}_5L\,^F_X \, / \, _5L\,^F_{X-5});$

$\infty N^F_{85}\,(t+5) = (5N^F_{80}\,(t) + \infty N^F_{80}\,(t)) \,.\, (T^F_{85} \, / \, T^F_{80});$

$_5B_x\,[t,\,t+5] = 5 \times {}_5F_x \times [\{{}_5N\,^F_X\,(t) + {}_5N\,^F_X\,(t+5)\} \, / \, 2]$

= births to women aged x to x + 5 between time f and time t + 5.

$$B^F\,[t,\,t+5] = \sum_{x=\alpha}^{\beta-5} 5Bx\,[t,\,t+5] \times \frac{1}{1} + 1.05$$

= number of females births between t and t + 5 (with SRB = 1.05).

$_5N_0{}^F(t+5) = B^F\,[t,\,t+5] \times 5\,L_0{}^F \, / \, 5l_0$

For Males:

$_5N\,^M_X\,(t)$ = number of women aged x to x+5 at time t.

$_5L\,^M_X$ = Number of person-years lived by women from age x to x + 5.

$$_5N\,^M{}_X\,(t+5) = {}_5N\,^M{}_{X-5}{}^{(t)} \times ({}_5L\,^M{}_X / {}_5L\,^M{}_{X-5});$$

$$\infty N^M{}_{85}\,(t+5) = (5N^M{}_{80}\,(t) + \infty N^M{}_{80}\,(t)) \,.\, (T^M{}_{85} / T^M{}_{80});$$

$$B^M\,[t, t+5] = \sum_{x=a}^{\beta-5} 5Bx\,[t, t+5] \times \frac{1}{1} + 1.05$$

= number of the males births between t and t + 5.

$$_5N_0{}^M(t+5) = B^M\,[t, t+5] \times 5\,L_0{}^M / 5l_0$$

****

Chapter – 4

World Demographic Profile

Following table shows the world demographic profile, according to the data of CIA World Factbook, 2017.

World Population	7,405,107,650 (July 2017)
10 Most Populated Countries (In Millions)	**China:** 1379.3 **India:** 1281.93 **United States:** 326.6 **Indonesia:** 260.58 **Brazil:** 207.35 **Pakistan:** 204.92 **Nigeria:** 190.63 **Bangladesh:** 157.83 **Russia:** 142.26 **Japan:** 126.45
Age Structure	**0-14 years:** 25.44% (male 963,981,944/female 898,974,458) **15-24 years:** 16.16% (male 611,311,930/female 572,229,547) **25-54 years:** 41.12% (male 1,522,999,578/female 1,488,011,505) **55-64 years:** 8.6% (male 307,262,939/female 322,668,546) **65 years and over:** 8.68% (male 283,540,918/female 352,206,092)
Dependency	**total dependency ratio:** 52.5

Ratio	**youth dependency ratio:** 39.9
	elderly dependency ratio: 12.6
	potential support ratio: 7.9
Median age	**total:** 30.4 years
	male: 29.6 years
	female: 31.1 years
Birth rate	4.3 births every second
Death rate	1.8 deaths every second
Maternal mortality	216 deaths/100,000 live births
Sex ratio	**at birth:** 1.03 male(s)/female
	0-14 years: 1.07 male(s)/female
	15-24 years: 1.07 male(s)/female
	25-54 years: 1.02 male(s)/female
	55-64 years: 0.95 male(s)/female
	65 years and over: 0.81 male(s)/female
	total population: 1.02 male(s)/female
Life Expectancy	**total population:** 69 years
	male: 67 years
	female: 71.1 years
Total Fertility Rate	2.42 children born/woman
Languages	**Mandarin Chinese:** 12.2%
	Spanish: 5.8%
	English: 4.6%
	Arabic: 3.6%
	Hindi: 3.6%
	Portuguese: 2.8%

Bengali: 2.6%

Russian: 2.3%

Japanese: 1.7%

- Percentages for "first language" speakers only; the six UN languages - Arabic, Chinese (Mandarin), English, French, Russian, and Spanish which are mother tongues for approximately half of the world's population. They are also the official languages in over half the countries in the world with more than a million first-language speakers.

- There is an estimated 7,000 languages spoken in the world and about 80% of the languages are spoken by >100,000 people and approximately 130 of those languages are spoken by >10 people.

- There is an estimated amount of 2,300 languages spoken in Asia, 2,140 in Africa, 1,310 in the Pacific, 1,060 in the Americas, and 290 in Europe.

Religions | **Christian:** 31.4%

Muslim: 23.2%

Hindu: 15%

Buddhist: 7.1%

folk religions: 5.9%

Jewish: 0.2%

other: 0.8%

unaffiliated: 16.4%

Source: CIA World Factbook, 2017

****

Chapter – 5

Demographic Profiles of the Populated Countries in the World

Following tables shows the demographic profiles of the populated Countries (most three) in the World, according to the data of CIA World Factbook, 2017.

China

Population	1,384,688,986
Age Structure	**0-14 years:** 17.15% (male 127,484,177/female 109,113,241)
	15-24 years: 12.78% (male 94,215,607 per female 82,050,623)
	25-54 years: 48.51% (male 341,466,438 per female 327,661,460)
	55-64 years: 10.75% (male 74,771,050 per female 73,441,177)
	65 years and over: 10.81% (male 71,103,029 per female 77,995,969)
Dependency Ratio	**total dependency ratio:** 37.7%
	youth dependency ratio: 24.3%
	elderly dependency ratio: 13.3%
	potential support ratio: 7.5%
Population Growth	0.41%
Death Rate	7.8 deaths per 1,000 people
Birth Rate	12.3 births per 1,000 people
Sex Ratio	**at birth:** 1.15 male(s) per female
	0-14 years: 1.17 male(s) per female
	15-24 years: 1.14 male(s) per female

	25-54 years: 1.04 male(s) per female
	55-64 years: 1.02 male(s) per female
	65 years and over: 0.92 male(s) per female
	total population: 1.06 male(s) per female
Maternal Mortality	27 deaths per 100,000 live births
Infant Mortality	**total:** 12 deaths per 1,000 live births
	male: 12.3 deaths per 1,000 live births
	female: 11.7 deaths per 1,000 live births
Life Expectancy	**Average:** 75.7 years
	male: 73.6 years
	female: 78 years
Total Fertility Rate	1.6 children born per woman
Ethnic Groups	**Han Chinese:** 91.6%
	Zhuang: 1.3%,
	other: 7.1% (Hui, Manchu, Uighur, Miao, Yi, Tujia, Tibetan, Mongol, etc.)
Religions	**Buddhist:** 18.2%
	Christian: 5.1%
	Muslim: 1.8%
	folk religion: 21.9%
	Hindu: < 0.1%
	Jewish: < 0.1%
	other: 0.7% (Daoist or Taoist)
	unaffiliated: 52.2%
Languages	- Standard Chinese or Mandarin
	- Yue (Cantonese)
	- Wu (Shanghainese)
	- Minbei (Fuzhou)
	- Minnan (Hokkien-Taiwanese)

	- Xiang
	- Gan
Literacy	**total population: 96.4%**
	male: 98.2%
	female: 94.5%

Source: CIA World Factbook, 2017

India

Population	1,281,935,911
Age Structure	**0-14 years:** 27.34% (male 186,087,665 per female 164,398,204)
	15-24 years: 17.9% (male 121,879,786 per female 107,583,437)
	25-54 years: 41.08% (male 271,744,709/female 254,834,569)
	55-64 years: 7.45% (male 47,846,122 per female 47,632,532)
	65 years and over: 6.24% (male 37,837,801 per female 42,091,086)
Dependency Ratio	**total dependency ratio: 52.2%**
	youth dependency ratio: 43.6%
	elderly dependency ratio: 8.6%
	potential support ratio: 11.7%
Population Growth	1.17%
Birth Rate	19 births per 1,000 people

Death Rate	7.3 deaths per 1,000 people
Sex Ratio	**at birth:** 1.12 male(s) per female
	0-14 years: 1.13 male(s) per female
	15-24 years: 1.13 male(s)/female
	25-54 years: 1.06 male(s)/female
	55-64 years: 1.01 male(s)/female
	65 years and over: 0.9 male(s)/female
	total population: 1.08 male(s)/female
Infant Mortality	**total:** 39.1 deaths per 1,000 live births
	male: 38 deaths per 1,000 live births
	female: 40.4 deaths per 1,000 live births
Life Expectancy	**total population:** 68.8 years
	male: 67.6 years
	female: 70.1 years
Total Fertility Rate	2.43 children born per woman
Maternal Mortality	174 deaths per 100,000 live births
Ethnic Groups	**Indo-Aryan:** 72%
	Dravidian: 25%
	Other: 3%
Religions	**Hindu:** 79.8%
	Muslim: 14.2%
	Christian: 2.3%
	Sikh: 1.7%
	other and unspecified: 2%
Languages	**Hindi:** 41%
	Bengali: 8.1%

	Telugu: 7.2%
	Marathi: 7%
	Tamil: 5.9%
	Urdu: 5%
	Gujarati: 4.5%
	Kannada: 3.7%
	Malayalam: 3.2%
	Oriya: 3.2%
	Punjabi: 2.8%
	Assamese: 1.3%
	Maithili: 1.2%
	other: 5.9%
Literacy	total population: 71.2%
	male: 81.3%
	female: 60.6%

Source: CIA World Factbook, 2017

The United States

Population	326,625,791
Age Structure	**0-14 years:** 18.73% (male 31,255,995 per female 29,919,938)
	15-24 years: 13.27% (male 22,213,952 per female 21,137,826)
	25-54 years: 39.45% (male 64,528,673 per female 64,334,499)
	55-64 years: 12.91% (male 20,357,880 per female

21,821,976)

65 years and over: 15.63% (male 22,678,235 per female 28,376,817)

Dependency Ratios	**total dependency ratio:** 51.2
	youth dependency ratio: 29
	elderly dependency ratio: 22.1
	potential support ratio: 4.5
Population Growth Rate	0.81%
Birth Rate	12.5 births per 1,000 people
Death Rate	8.2 deaths per 1,000 people
Net Migration	3.9 migrant(s) per 1,000 people
Sex Ratio	**0-14 years:** 1.04 male(s) per female
	15-24 years: 1.05 male(s) per female
	25-54 years: 1 male(s) per female
	55-64 years: 0.93 male(s) per female
	65 years and over: 0.79 male(s) per female
	total population: 0.97 male(s) per female
Infant Mortality	**total:** 5.8 deaths per 1,000 live births
	male: 6.3 deaths per 1,000 live births
	female: 5.3 deaths per 1,000 live births
Ethnic Groups	**White:** 72.4%
	Black: 12.6%
	Asian: 4.8%
	Amerindian and Alaska native: 0.9%
	native Hawaiian and other Pacific Islander: 0.2%
	Other: 6.2%
	'Two or more races: 2.9%

Maternal Mortality	14 deaths per 100,000 live births
Education Expenditure	4.9% of GDP
Languages	**English:** 79%
	Spanish: 13%
	other Indo-European: 3.7%
	Asian and Pacific island: 3.4%
	Other: 1%
Religions	**Protestant:** 46.5%
	Roman Catholic: 20.8%
	Jewish: 1.9%
	Mormon: 1.6%
	other Christian: 0.9%
	Muslim: 0.9%
	Jehovah's Witness: 0.8%
	Buddhist: 0.7%
	Hindu: 0.7%:
	other: 1.8%
	unaffiliated: 22.8%
	don't know/refused: 0.6%
Total Fertility Rate	1.87 children born per woman
Life Expectancy at Birth	**total population:** 80 years
	male: 77.7 years
	female: 82.2 years

Source: CIA World Factbook, 2017

****

Chapter – 6

Demography of New Zealand

The 2018 census enumerated a resident population of 4,699,755 – a 10.8 percent increase over the population recorded in the 2013 census. As of September 2020, the total population has risen to an estimated 5,030,360. In May 2020, Statistics New Zealand reported that New Zealand's population had climbed above 5 million people in March 2020.

The median child birthing age was 30 and the total fertility rate is 2.1 births per woman in 2010. In Māori populations, the median age is 26, and the fertility rate of 2.8. In 2010 the age-standardized mortality rate was 3.8 deaths per 1000 (down from 4.8 in 2000) and the infant mortality rate for the total population was 5.1 deaths per 1000 live births. The life expectancy of a New Zealand child born in 2014-16 was 83.4 years for females and 79.9 years for males, which is among the highest in the world. Life expectancy at birth is forecast to increase from 80 years to 85 years in 2050 and infant mortality is expected to decline. In 2050 the median age is forecast to rise from 36 years to 43 years and the percentage of people 60 years of age and older rising from 18 percent to 29 percent. During early migration in 1858, New Zealand had 131 males for every 100 females, but following changes in migration patterns and the modern longevity advantage of women, females came to outnumber males in 1971. As of 2012, there are 0.99 males per female, with males dominating under 15 years and females dominating in the 65 years or older range.

Following case study shows the detailed demographic situations of New Zealand.

Data availability, data quality and demographic scenario of New Zealand

[1]Kousik Das Malakar

Abstract:

Aims and Objectives:

A demographic scenario is a form of demographic analysis used by social sciences research and allied fields so that they may be as efficient as possible with services and identifying any possible gaps in their society, planning, and others. The study seeks to examine that, to explore the data characteristics (data availability, data quality) and demographic scenario of New Zealand.

Data and Methods:

In this study, data were taken from the New Zealand census 2018 (Statistics New Zealand, 2018). The data is analysed in two modes, i.e. statistical and cartographic analysis, with the help of MS Excel 2013, ArcGIS 10.2.2 software. And using the equations of Population growth rate, YADR (Young Age Dependency Ratio), OADR (Old-Age Dependency Ratio), Sex Ratio, Birth rate, Death rate, Total fertility rate, Natural change between Crude birth rate and Crude death rate, Net migration, Mother's mean age at first birth, Maternal mortality rate, Infant mortality rate, Life expectancy at birth, Population projections, Unemployment Rate (U), and others.

Results and conclusion:

From this discussion, the country of New Zealand is a rich country and living way is good and GDP is \$199 billion and per capita income is \$40,266, and the HDI value is very high 0.921. And the social conditions are not so bad. Here, presents the unity in diversity. And the child and youth stages population are not good conditions (increasing rate is low), it will be affected the future generation.

Keywords: Demographic Profile; Data Characteristics; Social demography; New Zealand

---

[1]Centre for the Study of Regional Development, School of Social Sciences, Jawaharlal Nehru University (JNU), New Delhi, India. E-mail: kdmalakargegr92@gmail.com

1. Introduction:

A demographic scenario is a form of demographic analysis used by social sciences research and allied fields so that they may be as efficient as possible with services and identifying any possible gaps in their society, planning, and others. The word "demographics" comes from the Ancient Greek: "demo" meaning people and "graphics" meaning measurement. There is a strong tradition of studying demography as part of economics (Mester, 2017). Demographics are the quantifiable statistics of a given population. Demographics are also used to identify the study of quantifiable subsets within a given population which characterize that population at a specific point in time. Demographic profiling is essentially an exercise in making generalizations about groups of people. As with all such generalizations many individuals within these groups will not conform to the profile - demographic information is aggregate and probabilistic information about groups, not about specific individuals. Demographics include such factors as gender, age, ethnicity, occupation, seniority, salary levels, marital and family status.

The demographics of New Zealand encompass the gender, ethnic, religious, geographic, and economic backgrounds of the 5 million people living in New Zealand. According to 2018 census (Statistics New Zealand), the majority of New Zealand's population is of European descent (70 percent), with the indigenous Māori being the largest minority (16.5 percent), followed by Asians (15.3 percent), and non-Māori Pacific Islanders (9.0 percent). This is reflected in immigration, with most new migrants coming from Britain and Ireland, although the numbers from Asia, in particular, are increasing. Auckland is the most ethnically diverse region in New Zealand with 43.0 percent identifying as Europeans, 28.5 percent as Asian, 11 percent as Māori, 15.5 percent as Pacific Islanders, and 2 percent as Middle Eastern, Latin American or African (MELAA). Compared to the diversity of the population as a whole, the population aged under 18 years is considerably more ethnically diverse.

2. Aims and Objectives:

2.1 Aims:

This paper aims to explore the data characteristics (data availability, data quality) and demographic scenario of New Zealand.

2.2 Objectives:

The specific objectives of the study are;

2.2.1 To understand the data characteristics especially; data availability, data quality.

2.2.2 To present a demographic scenario of New Zealand.

3. Study area:

3.1 Terminology: The name derives from the 'kiwi', a native flightless bird, which is the national symbol of New Zealand. The Māori loanword 'Pakeha' usually refers to New Zealanders of European descent, although some reject this appellation, and some Maori uses it to refer to all non-Polynesian New Zealanders.

3.2 Location: New Zealand (fig. 1) is an island country in the southwestern Pacific Ocean. It comprises two main landmasses—the North Island and the South Island and around 600 smaller islands, covering a total area of 268,021 square kilometres. New Zealand is about 2,000 kilometres east of Australia across the Tasman Sea and 1,000 kilometres south of the islands of New Caledonia, Fiji, and Tonga. The country's varied topography and sharp mountain peaks, including the Southern Alps, owe much to tectonic uplift and volcanic eruptions. New Zealand's capital city is Wellington, and its most populous city is Auckland.

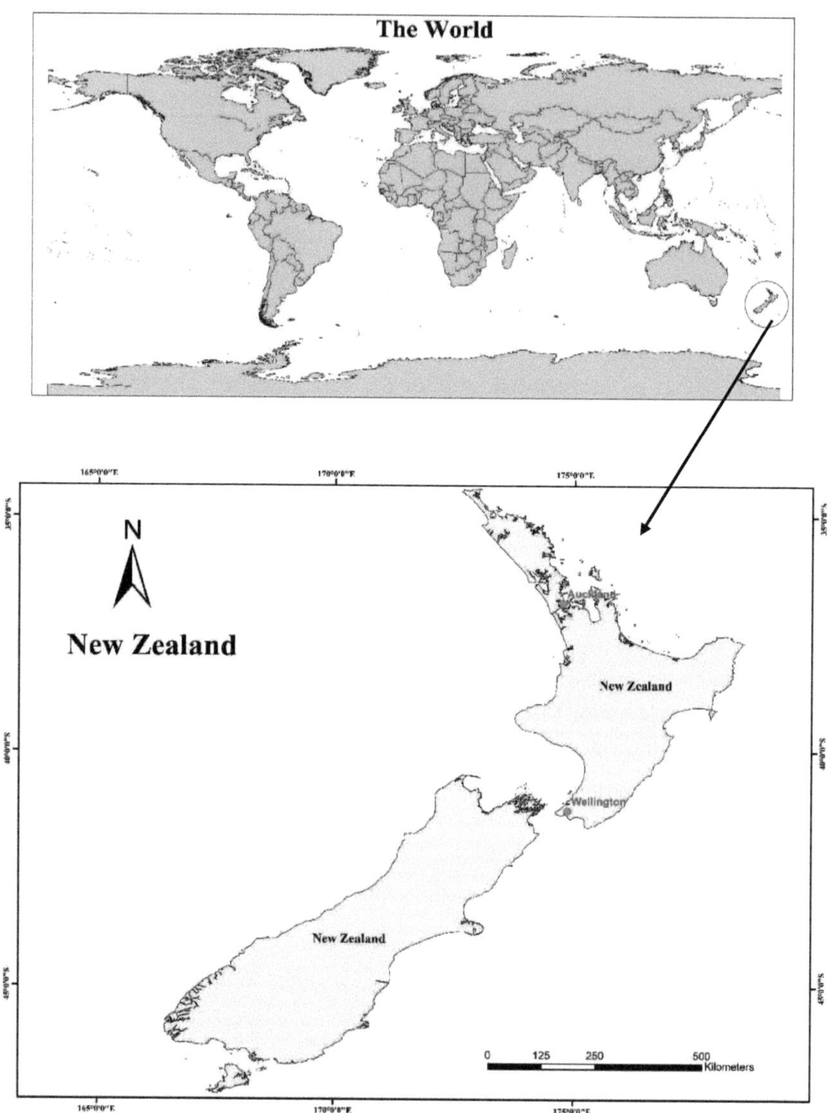

Figure 2: Study area Map

Following table 1. Shows the an overviews concept of New Zealand

Table 1. **New Zealand**

Capital	Wellington	**Largest city**	Auckland
Official languages	English, Maori, NZ Sign Language	**Ethnic groups (2018)**	71.8% European; 16.5% Māori; 15.3% Asian; 9.0% Pacific peoples; 1.5% ME/LA/African; 1.2% Other
Religion (2018)	48.5% No religion; 37.0% Christianity; 2.6% Hinduism; 1.3% Islam; 1.1% Buddhism; 1.9% Others; 6.6% Undeclared	**Demonym(s)**	New Zealander Kiwi (informal)
Government **Unitary parliamentary constitutional monarchy**	Monarch: Elizabeth II Governor-General: Patsy Reddy Prime Minister: Jacinda Ardern	**Legislature**	Parliament (House of Representatives)
Stages of independence from the United Kingdom	Responsible government: 7 May 1856 Dominion: 26 September 1907 Statute of Westminster adopted: 25 November 1947	**Area**	Total: 268,021 km2 Water (%): 1.6
Population	August 2020 estimate: 5,028,980 (120th). 2018 census: 4,699,755. Density: 18.3/km2 (47.4/sq mi) (167th).	**GDP (PPP) 2018 estimate**	Total: $199 billion Per capita: $40,266
GDP (nominal) 2018 estimate	Total: $206 billion Per capita: $41,616	**Gini (2019)**	Negative increase 33.9
HDI (2018)	Increase 0.921: very high.	**Currency**	New Zealand dollar ($) (NZD)
Time zone	UTC+12 Summer (DST) UTC+13	**Source:** Statistics New Zealand, 2018	

4. Data Characteristics and sources of data:

4.1 Data availability:

Data availability is a term used by some computer storage manufacturers and storage service providers (SSPs) to describe products and services that ensure that data continues to be available at a required level of performance in situations ranging from normal through "disastrous." In general, data availability is achieved through redundancy involving where the data is stored and how it can be reached. Some vendors describe the need to have a data center and a storage-centric rather than a server-centric philosophy and environment.

Data availability is about the timeliness and reliability of the access to and use of data. It includes data accessibility. Availability has to do with the accessibility and continuity of information. Information with low availability concerns may be considered supplementary rather than necessary. Information with high availability concerns is considered critical and must be accessible in order to prevent a negative impact on University activities.

Examples of data with high availability concerns include:

- Website files, which must remain accessible to prevent site downtime and disruption of service.
- Payroll and tuition data, which must remain accessible to facilitate business continuity.

Consider the following when managing data availability:

- Whether data needs to remain accessible;
- Which storage methods and locations are suitably accessible;
- Weather data is easily re-acquired or re-created if lost, stolen, or destroyed;
- Data integrity and data availability are both factors in data's criticality, or how essential that data is to the University's operations.

An important thinks of data availability is Confidentiality, Integrity and Availability Model

Confidentiality, integrity, and availability (also known as the CIA triad) is a model designed to help organizations plan their information security strategy and comply with data protection regulations.

- **Confidentiality**: A set of rules and procedures to limit unauthorized access to sensitive information. This includes measures such as training security teams to identify vulnerabilities across the organization's environment, training employees to avoid data misuse, and implementing the use of strong passwords.
- **Integrity:** Ensuring the accuracy, consistency, and reliability of data. Security teams must take steps to ensure the integrity of data at rest and in transit. Protective measures against attacks that can modify data include file permissions, user access controls and version controls.
- **Availability:** The ability to guarantee reliable access to data. Organizations must keep crucial data available and shorten data outage times as much as possible. To achieve data availability, organizations must be able to quickly repair all hardware failures and maintain backups.

Data Availability Challenges

There are several issues that can affect the availability of your data:

- Host server failures: If the server that stores your data fails, your data will become unavailable.
- Storage failures: If your physical storage device fails, you can no longer access the data it stores.
- Network crash: If the network crashes, the host server becomes inaccessible along with the data stored on it.
- Poor data quality: Low-quality datasets may contain incomplete, inconsistent, or redundant data, which could be useless for your IT operations.
- Data compatibility issues: Data that is usable and working on a specific platform or environment might not be on another.
- Legacy data: Data that is too outdated can become unusable. You can use data transformation tools to make older data readily accessible, but these do not always work.

Best Practices and Tools to Ensure Data Availability:

Data Redundancy

It is important to store backups of the data in a separate location, or in a distributed network. This ensures that if a storage component degrades or fails, won't lose the data permanently.

Data Loss Prevention Tools

Data Loss Prevention (DLP) tools help mitigate against data breaches and physical damage to the data center. They leverage cloud-based or third-party secure storage to prevent data loss. Some DLP tools include features such as monitoring, threat blocking, and forensic analysis.

Erasure Coding

Object storage uses advanced erasure coding to ensure that data is always available. Erasure coding combines data with parity information and then splits or "shards" it and distributes it across the storage environment.

Guidelines for data availability

- Inventory data.
- Securely dispose of data, devices, and paper records.
- Use official University accounts and systems rather than personal ones.

4.2 Data quality:

Data quality refers to the state of qualitative or quantitative pieces of information. There are many definitions of data quality, but data is generally considered high quality if it is "fit for intended uses in operations, decision making, and planning". Moreover, data is deemed of high quality if it correctly represents the real-world construct to which it refers.

The main purpose of ensuring data quality in Extension research and evaluation studies is to present information that is credible. Such research and evaluation studies follow research protocols, conducted in an ethical manner, and withstand the test of scrutiny by reviewers. Data quality is generally understood to be the degree to which data, including research processes such as data collection and statistical accuracy, meet the needs of users. Among the critical aspects to consider when assessing data for quality are relevance, validity, reliability, objectivity, integrity, completeness, generalizability, and utility. Ensuring these critical aspects of data quality in Extension research and evaluation studies is of paramount importance if Extension is to implement and improve programming based on sound methods.

Defining data quality in a sentence is difficult due to the many contexts data are used in, as well as the varying perspectives among end-users, producers, and custodians of data. There are such as;

Business perspective, data quality is:

+ Data that is "'fit for use' in their intended operational, decision-making and other roles" or that exhibits "'conformance to standards' that have been set so that fitness for use is achieved."
+ Data that "are fit for their intended uses in operations, decision making, and planning."
+ "the capability of data to satisfy the stated business, system, and technical requirements of an enterprise."

Consumer perspective, data quality is:

+ "data that are fit for use by data consumers"
+ data "meeting or exceeding consumer expectations"
+ data that "satisfies the requirements of its intended use"

Standards-based perspective, data quality is:

+ the "degree to which a set of inherent characteristics of an object fulfils requirements."
+ "the usefulness, accuracy, and correctness of data for its application."

The various nature of data quality, there are such as,

+ accessibility or availability;
+ comparability;
+ accuracy or correctness;
+ consistency, coherence, or clarity;
+ completeness or comprehensiveness;
+ uniqueness;
+ validity or reasonableness;
+ credibility, reliability, or reputation;
+ relevance, pertinence, or usefulness;
+ timeliness or latency.

Data quality assurance

Data quality assurance is the process of data profiling to discover inconsistencies and other anomalies in the data, as well as performing data cleansing activities (like an example, removing outliers, missing data interpolation) to improve the data quality.

These activities can be undertaken as part of data warehousing or as part of the database administration of an existing piece of application software.

Data quality control

Data quality control is the process of controlling the usage of data for an application or a process. This process is performed both before and after a Data Quality Assurance (QA) process, which consists of the discovery of data inconsistency and correction.

Before: Restricts inputs

After: QA process the following statistics are gathered to guide the Quality Control (QC) process:

- Severity of inconsistency
- Incompleteness
- Accuracy
- Precision
- Missing / Unknown

The Data QC process uses the information from the QA process to decide to use the data for analysis or in an application or business process. General example: if a Data QC process finds that the data contains too many errors or inconsistencies, then it prevents that data from being used for its intended process which could cause disruption. Specific example: providing invalid measurements from several sensors to the automatic pilot feature on an aircraft could cause it to crash. Thus, establishing a QC process provides data usage protection.

Eight Components of Data Quality:

Validity

Validity refers to the "closeness between the values provided and the true values" (Organization for Economic Cooperation and Development The careful development of the questionnaire provides a basis for validity. A thorough examination of previous studies, an ongoing review by a panel of experts, and carrying out a field test makes the case for construct, content, and face validity.

Reliability

Reliability is determined by the degree to which measurements are similar (consistent) on repeated measurements. The careful wording of the questionnaire and pilot testing the questionnaire with subjects not included in the sample, as well as a high response rate, provide evidence for reliability.

Objectivity

The objectivity of data means that conclusions are based on statistically sound methods. Careful analysis of assumptions/hypotheses/objectives/research questions and use of appropriate statistical procedures and results provide evidence of objectivity.

Integrity

Integrity is concerned with minimizing errors through the process of collecting, recording, and analyzing data. Integrity can be enhanced by properly training those involved with data collection and by reviewing that the data have been properly recorded.

Generalizability

Generalizability is concerned with sound sampling procedures that yield a sample representative of the population on key variables and follow-up with non-respondents.

Completeness

Completeness refers to ways in which missing values that exist in a given dataset are handled. When data are missing at random, their incompleteness is due to external events that cannot be controlled, whereas data not missing at random cannot be collected due to known and expected external events. The data not missing at random must be considered during data analysis to better understand the limitations and generalizability of the study.

Relevance

Relevance refers to the degree to which data are important to users and their needs. Among the strategies to ensure a high degree of relevancy are thorough literature reviews and needs assessments.

Utility

The utility includes aspects of timeliness (data collected in a timely manner so that data maintain their relevance to their users), punctuality (release of data), and accessibility (ways in which data are made available to the intended users).

Example of data quality in public health:

Data quality has become a major focus of public health programs in recent years, especially as demand for accountability increases. Work towards ambitious goals related to the fight against diseases such as AIDS, Tuberculosis, and Malaria must be predicated on strong Monitoring and Evaluation systems that produce quality data related to program implementation. These programs, and program auditors, increasingly seek tools to standardize and streamline the process of determining the quality of data, verify the quality of reported data, and assess the underlying data management and reporting systems for indicators. An example is WHO and MEASURE Evaluation's Data Quality Review Tool WHO, the Global Fund, GAVI, and MEASURE Evaluation have collaborated to produce a harmonized approach to data quality assurance across different diseases and programs

4.3 Sources of data:

In this study, data were taken from the New Zealand census 2018 (Statistics New Zealand, 2018).

5. Methodology:

Overall, Methodology shows the table number 2.

Table 2: Mathematical Expression / Methods / Equations

Sl_No	Name	Mathematical Expression / Methods / Equations	Remarks
1	**Population growth rate (G)**	$G = [\{(P_1 - P_2)/P_1\} * 100]$	G- Population growth rate. P_1- Population of the base year. P_2- Population of the present year.
2	**YADR (Young Age Dependency Ratio)**	[(Total population in 0 to 14 years age/ Total population in 15 to 59 years age)*100].	
3	**OADR (Old Age Dependency Ratio)**	[(Total population in 60+ years age/ Total population in 15 to 59 years age)*100].	Used the value of 'K'

4	**Sex Ratio**	(Male populations/Female populations) × K (1000 or 100)	is based on Developed and Developed countries number of populations
5	**Birth rate**	(Number of Births / Total Populations) × 1000	
6	**Death rate**	(Number of deaths / Total Populations) × 1000	
7	**Total fertility rate**	This is an overall summary measure of fertility and is obtained by summing the age-specific fertility rate for each age of the childbearing span. In other words, the total fertility rate is the number of children that a woman of the hypothetical cohort would bear during her lifetime if she were to bear children throughout her life at the Age-Specific Fertility Rates for a given year and if none of them dies before crossing the age of reproduction. It is constructed as follows; TFR = $\sum f_x$ [$\sum$ (five-year age-specific birth rates for females aged 10 to 49) × 5] Therefore, the **age Specific Marital Fertility Rate (ASMFR):** ASMFR is measured as number of births per year in a given age group to the total number of married women in that age group at mid-year. It is constructed as follows; $$ASMFR = \frac{nBx}{nWmx}$$	

		Where, 'nBx' is the number of births in a year to the married women of ages x to x+n years in a given years and geographical region; '$_nW^mx$' is the number of women aged x to x+n years at mid-year in a given year and geographical area.	
8	**Natural change between Crude birth rate and Crude death rate**	(Crude birth rate - Crude death rate)	
9	**Net migration**	$= \frac{I - O}{P} \times 1000$ Where, ❖ Rate of In-migration (Im) = $\frac{I}{P} \times 1000$ ❖ Rate of Out-migration (Em) $= \frac{O}{P} \times 1000$	I- In-migration; O- Out-migration; P- Place of Origin or Place of destination.
10	**Mother's mean age at first birth**	$M(t) = \frac{\sum_0^{amax}(a+0.5)\,b\,(a,t)}{\sum_0^{amax} b\,(a,\ t)}$	Where, M(t) = Average age at first birth at time t b(a,t) = the age-specific birth rate for birth order one at (single) age a and time t. a_{max}= the highest age at which first births are

observed.

11	**Maternal mortality rate**	(Number of resident maternal deaths/Number of resident live births) x 100,000
12	**Infant mortality rate**	[(number of deaths to live born infants under one year of age) / (number of births)] × 1000 Based on life table estimation, **Observed data:** $_nN_x$ = mid-year population in age interval x to x +n $_nD_x$ = deaths between ages x and x +n during the year
13	**Life expectancy at birth**	**Steps for period life table construction:** 1. $_nm_x \approx nM_x = (_nD_x / _nN_x)$. 2. $_na_x$: calculated from Coale and Demeny equations. 3. $_nq_x = \{(n \times _nm_x) / (1 + (n - _na_x) \times _nm_x)\}$ $_\infty q_{85} = 1.00$ 4. $_nP_x = 1 - _nq_x$ 5. $l_0 = 100,000$ $l_{x+n} = l_x \cdot _nP_x$ 6. $_nd_x = l_x - l_{x+n}$ 7. $_nL_x = n \cdot l_{x+n} + _na_x \cdot _nd_x$ 8. $T_x = \sum_{a=x}^{\omega} nLa$ 9. $e^0_x = T_x / l_x$
14	**Population projections**	**Cohort-component method:** Cohort-component method is widely used method for the population projections. Following the description of this method; **For female:** $_5N^F_x (t)$ = number of women aged x to x+5 at time t. $_5L^F_x$ = Number of person-years lived by women from age x to x + 5. $_5F_x$ = age-specific fertility rate in interval x to x + 5. $_5N^F_x (t+5) = _5N^F_{x-5}{}^{(t)} \times (_5L^F_x / _5L^F_{x-5})$;

$\infty N^F_{85}(t+5) = (5N^F_{80}(t) + \infty N^F_{80}(t)) \cdot (T^F_{85} / T^F_{80});$

$_5B_x[t, t+5] = 5 \times {_5F_x} \times [\{_5N^F_x(t) + {_5N^F_x}(t+5)\} / 2]$

= births to women aged x to x + 5 between time f and time t + 5.

$B^F[t, t+5] = \sum_{x=\alpha}^{\beta-5} 5Bx[t, t+5] \times \frac{1}{1} + 1.05$

= number of females births between t and t + 5 (with SRB = 1.05).

$_5N_0^F(t+5) = B^F[t, t+5] \times 5\,L_0^F / 5l_0$

For Males:

$_5N^M_x(t) =$ number of women aged x to x+5 at time t.

$_5L^M_x =$ Number of person-years lived by women from age x to x + 5.

$_5N^M_x(t+5) = {_5N^M_{x-5}}^{(t)} \times ({_5L^M_x} / {_5L^M_{x-5}});$

$\infty N^M_{85}(t+5) = (5N^M_{80}(t) + \infty N^M_{80}(t)) \cdot (T^M_{85} / T^M_{80});$

$B^M[t, t+5] = \sum_{x=\alpha}^{\beta-5} 5Bx[t, t+5] \times \frac{1}{1} + 1.05$

= number of the males births between t and t + 5.

$_5N_0^M(t+5) = B^M[t, t+5] \times 5\,L_0^M / 5l_0$

15	**Calculation of percentage (i e, Religious, languages and other calculation for percentages)**	= (Individual values / Total Values) × 100

To calculate the unemployment rate, the number of unemployed people is divided by the number of people in the labour force,

| 16 | **Unemployment Rate (U)** | which consists of all employed and unemployed people. The ratio is expressed as a percentage.

$$U = \frac{\text{the number of unemployed peoples}}{\text{the number of people in the labour force}} \times 100$$ |
| 17 | **Others** | Some calculated data are direct collected from New Zealand data census Report.
There are such as;
Obesity - adult prevalence rate, Hospital bed density, Physician's density, Health expenditures, HIV/AIDS – deaths, HIV/AIDS - people living with HIV/AIDS, HIV/AIDS - adult prevalence rate |

6. Demographic scenario of New Zealand

Demography is the quantitative study of populations. Demographic data, in their simplest form, refer to six interacting dimensions: Births, deaths, migration (and resulting population growth), Age, sex, spatial distribution (and resulting population structure). Following sections are describe the various parts demographic scenario of New Zealand.

6.1 Population distribution:

The 2018 census enumerated a resident population of 4,939,755, a 10.8 percent increase over the population recorded in the 2013 census. As of August 2020, the total population has risen to an estimated 5,028,900. In May 2020, Statistics New Zealand reported that New Zealand's population had climbed above 5 million people in March 2020.

According to the Census of New Zealand (Statistics New Zealand, 2013, 2018), the median child birthing age was 30 and the total fertility rate is 2.1 births per woman in 2010. In Maori populations the median age is 26 and fertility rate 2.8. In 2010 the age-standardised mortality rate was 3.8 deaths per 1000 (down from 4.8 in 2000) and the infant mortality rate for the total population was 5.1 deaths per 1000 live births. The life expectancy of a New Zealand child born in 2014-16 was 83.4 years for females, and 79.9 years for males, which is among the highest in the world. Life expectancy at birth is forecast to increase from 80 years to 85 years in 2050 and infant mortality is expected to decline. In 2050 the median age is forecast to rise from 36 years to 43 years and the percentage of people 60 years of age and older rising from 18 percent to 29 percent. During early migration in 1858, New Zealand had 131 males for every 100 females, but following changes in migration patterns and the modern longevity advantage of women, females came to outnumber males in 1971. As of 2012 there are 0.99 males per female, with males dominating under 15 years and females dominating in the 65 years or older range. Following table 3 and figure 2 shows the population distribution of New Zealand.

Table. 3: Population Distribution of New Zealand

Sl_No	Years	Population (Millions)
1	1996	3.68
2	2001	3.82
3	2006	4.14
4	2014	4.56
5	2018	4.93

Source: Statistics New Zealand

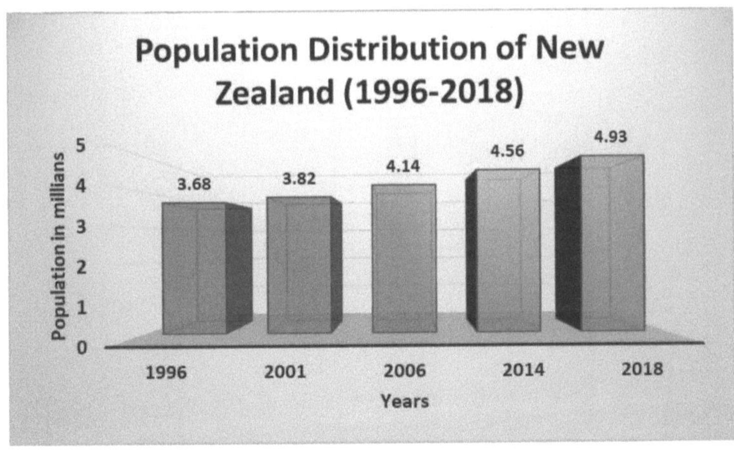

Figure 2: Population distribution of New Zealand (1996 - 2018)

6.2 New Zealand Population Growth:

New Zealand publishes its own online population clock and in late 2016, the estimated figure is shown as 4,723,562. Government statistics draw on their estimated levels of natural growth, and are slightly higher than the UN's estimates.

The New Zealand population clock projects from the June 2016 estimated population using the following factors:

- One birth every nine minutes
- One death every fifteen minutes
- One net migration gain every six minutes and twenty-nine seconds

New Zealand was actually losing citizens to migration as recently as 2012, but now has a significant positive net migration. This, coupled with the difference between births and deaths, contributes to a healthy rise in population that is expected to continue throughout the 21st century. The most recent New Zealand census was taken in 2018. The census is usually held every five years, but in 2011 New Zealand suffered from a major earthquake, which pushed the 2011 census back to 2013. According to the New Zealand Census, the population growth rate are figurate in figure 3.

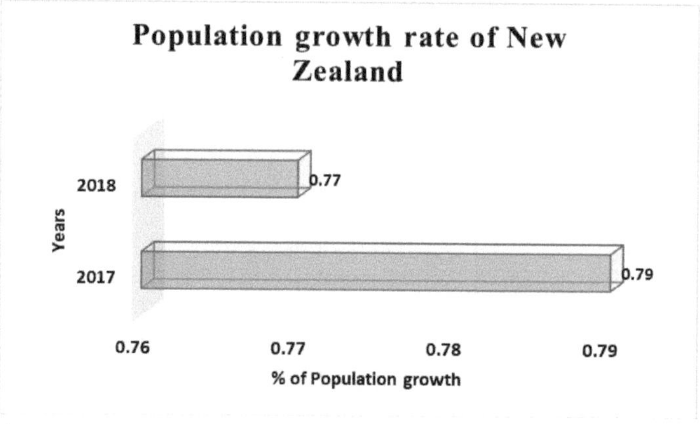

Figure 3: Population growth rate of New Zealand

70

6.3 Population density of New Zealand:

New Zealand's population density is relatively low, at 18 persons per square kilometre. The vast majority of the population live on the main North and South Islands, with New Zealand's major inhabited smaller islands being Waiheke Island (9,330), the Chatham and Pitt Islands (710), and Stewart Island (381). Over three-quarters of the population live in the North Island (76.5 percent), with one-third of the total population living in the Auckland Region. This region is also the fastest growing, accounting for 46 percent of New Zealand's total population growth. Most Māori live in the North Island (86.0 percent), although less than a quarter (23.8 percent) live in Auckland. New Zealand is a predominantly urban country, with 83.6 percent of the population living in an urban area. About 66.0 percent of the population live in the 20 main urban areas (population of 30,000 or more) and 45.3 percent live in the four largest cities of Auckland, Christchurch, Wellington, and Hamilton.

Approximately 14 percent of the population live in four different categories of rural areas as defined by Statistics New Zealand. About 18 percent of the rural population live in areas that have a high urban influence (roughly 12.9 people per square kilometre), many working in the main urban area. Rural areas with moderate urban influence and a population density of about 6.5 people per square kilometre account for 26 percent of the rural population. Areas with low urban influence where the majority of the residents work in the rural area house approximately 42 percent of the rural population. Remote rural areas with a density of less than 1 person per square kilometre account for about 14 percent of the rural population.

Population density of New Zealand [Persons per sq, km (I shown this value in CGS system)]

$$= \frac{4939755}{268021}$$

$= 18.43 = 18$ (Approx) Persons per sq, km.

6.4 Age-sex structure of New Zealand:

The composition of a population as determined by the number or proportion of males and females in each age category. The age-sex structure of a population is the cumulative result of past trends in fertility, mortality, and migration. Information on age-sex composition is essential for the description and analysis of many other types of demographic data.

A population pyramid illustrates (fig. 4) the age and sex structure of a country's population and may provide insights about political and social stability, as well as economic development. The population is distributed along the horizontal axis, with males shown on the right and females on the left. The male and female populations are broken down some selected age groups represented as horizontal bars along the vertical axis, with the youngest age groups at the bottom and the oldest at the top. The shape of the population pyramid gradually evolves over time based on fertility, mortality, and international migration trends.

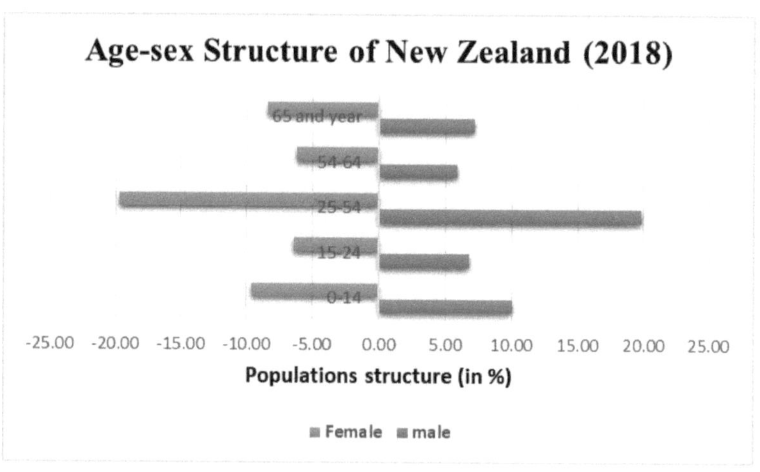

Figure 4: Age-sex structure of New Zealand

6.5 Dependency ratio:

In New Zealand, the total dependency ration is 52.9, youth dependency ratio is 30.5, elderly dependency ratio is 22.4, and the potential support ratio is 4.5.

6.6 Median age of the country of New Zealand:

In 2018 the median age of the New Zealand population was 38.1 years (male: 37.2 years and female: 39 years). By 2050 the median age is projected to rise to 43 years and the percentage of people 60 years of age and older to rise from 18% to 29%.

6.7 Sex-ratio:

At birth: 1.05 male(s)/female

0-14 years: 1.05 male(s)/female

15-24 years: 1.06 male(s)/female

25-54 years: 1.01 male(s)/female

55-64 years: 0.95 male(s)/female

65 years and over: 0.86 male(s)/female

Total population: 0.99 male(s)/female (2018 est.)

This entry includes the number of males for each female in five age groups - at birth, under 15 years, 15-64 years, 65 years and over, and for the total population. Sex ratio at birth has recently emerged as an indicator of certain kinds of sex discrimination in some countries. For instance, high sex ratios at birth in some Asian countries are now attributed to sex-selective abortion and infanticide due to a strong preference for sons. This will affect future marriage patterns and fertility patterns. Eventually, it could cause unrest among young adult males who are unable to find partners.

6.8 Birth rate and Death rate:

The ratio between births and individuals in a specified population and time. The data of 2017 shows that the birth rate of New Zealand is 13.2 births/1,000 populations and 13.1 births/1,000 populations are the latest census data for birth rate.

And the ratio between deaths and individuals in a specified population and time. The data of 2017 shows that the death rate of New Zealand is 7.5 deaths/1,000 populations and 7.6 deaths/1,000 populations are the latest census data for the death rate. Following figure 5 shows the situations of birth and death rate of New Zealand.

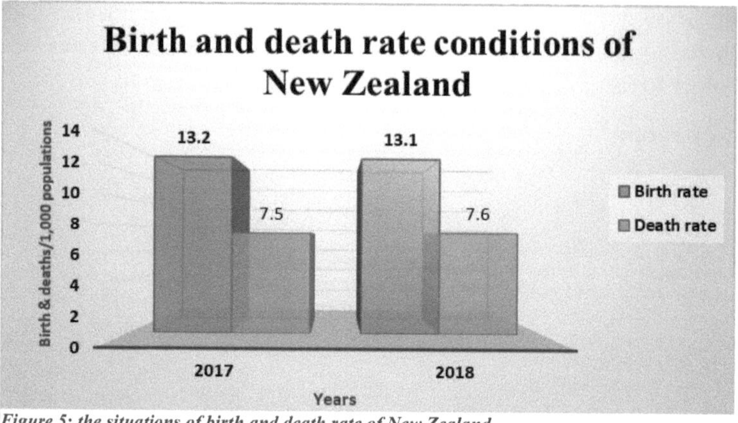

Figure 5: the situations of birth and death rate of New Zealand

6.9 Crude birth rate, Crude death rate, Natural change, and Total fertility rate of New Zealand:

The crude birth rate is the number of live births occurring among the population of a given geographical area during a given year, per 1,000 mid-year total population of the given geographical area during the same year. The crude death rate is the number of deaths occurring among the population of a given geographical area during a given year, per 1,000 mid-year total population of the given geographical area during the same year. And the total fertility rate is this is an overall summary measure of fertility and is obtained by summing the age-specific fertility rate for each age of the childbearing span. In other words, the total fertility rate is the number of children that a woman of the hypothetical cohort would bear during her lifetime if she were to bear children throughout her life at the Age-Specific Fertility Rates for a given year and if none of them dies before crossing the age of reproduction.

Following table 4 shows the overall things and figurate it (fig. 6 & 7).

Table 4. : Vital Statistics Since 2000

YEARS	Population	Live births	Deaths	Natural change	Crude birth rate (per 1,000)	Crude death rate (per 1,000)	Natural change (per 1,000)	Total fertility rate
2000	38,51,200	56,605	26,660	29,945	14.7	6.9	7.8	1.98
2001	38,86,700	55,800	27,825	27,972	14.36	7.16	7.2	1.97
2002	39,51,200	54,021	28,065	25,956	13.67	7.1	6.57	1.89
2003	40,27,700	56,136	28,011	28,125	13.94	6.95	6.99	1.93
2004	40,88,700	58,074	28,419	29,655	14.2	6.95	7.25	1.98
2005	41,36,000	57,744	27,033	30,711	13.96	6.54	7.42	1.97
2006	41,85,300	59,193	28,245	30,948	14.1	6.75	7.39	2.01
2007	42,26,200	64,044	28,521	35,520	15.15	6.75	8.4	2.18
2008	42,62,000	64,341	29,187	35,154	15.1	6.85	8.25	2.19
2009	43,04,900	62,541	28,965	33,579	14.53	6.73	7.8	2.13
2010	43,53,000	63,897	28,437	35,457	14.68	6.53	8.15	2.17
2011	43,86,300	61,404	30,081	31,320	14	6.86	7.14	2.09
2012	44,10,700	61,179	30,099	31,080	13.87	6.82	7.05	2.1
2013	44,46,700	58,719	29,568	29,148	13.2	6.65	6.55	2.01
2014	45,13,200	57,243	31,062	26,181	12.68	6.88	5.8	1.92
2015	45,99,300	61,038	31,608	29,430	13.27	6.87	6.4	1.99
2016	46,96,500	59,43	31,17	28,251	12.6	6.64	6.01	1.87

2017	47,96,000	59,610	33,339	26,268	12.43	6.95	5.48	1.81
2018	49,29,700	58,020	33,225	24,795	11.76	6.73	5.03	1.71

Source: Statistics New Zealand

6.10 Net migration rate and Urbanization:

The net migration rate is the difference between the number of immigrants (people coming into an area) and the number of emigrants (people leaving an area) throughout the year. When the number of immigrants is larger than the number of emigrants, a positive net migration rate occurs. A positive net migration rates indicate that there are more people entering than leaving

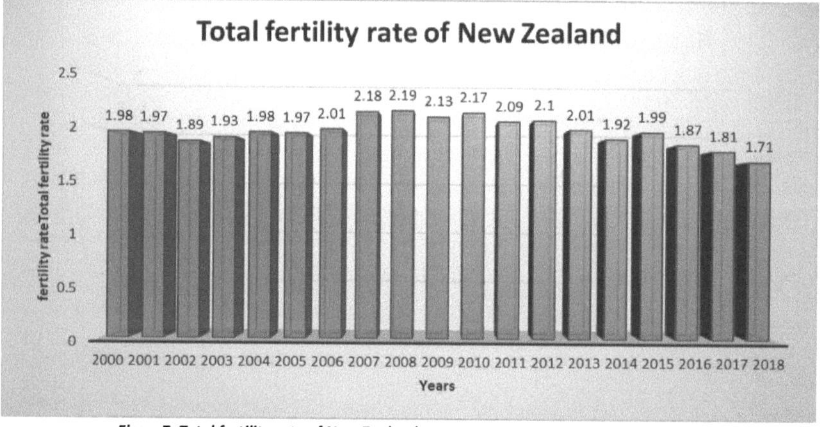

Figure7: Total fertility rate of New Zealand

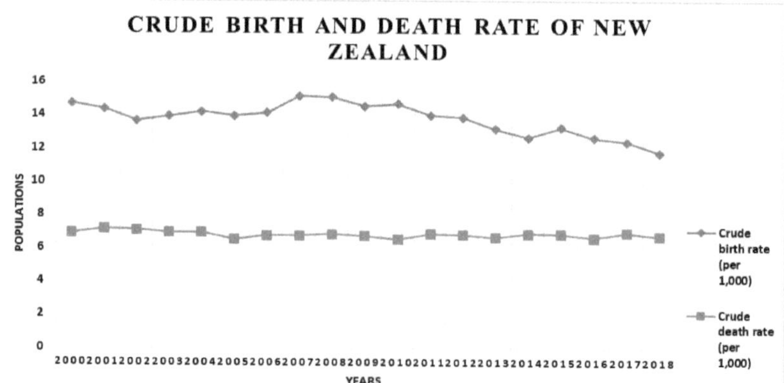

an area.

When more emigrate from a country, the result is a negative net migration rate, meaning that more people are leaving than entering the area. When there is an equal number of immigrants and emigrants, the net migration rate is balanced. The net migration rate of New Zealand is 2.2 migrant(s)/1,000 populations. And the urbanization rate is 86.6% of total populations and rate of urbanization is 1.01% annual rate of change.

6.11 Mother's mean age at first birth, maternal mortality rate, and Infant mortality rate

In New Zealand, the mother's mean age at first birth is 27.8 years, the maternal mortality rate is 9 deaths/100,000 live births and the Infant mortality rate is total: 4.4 deaths/1,000 live births; male: 4.9 deaths/1,000 live births; female: 3.8 deaths/1,000 live births.

6.12 Obesity - adult prevalence rate, Hospital bed density, Physician's density, Health expenditures, HIV/AIDS – deaths, HIV/AIDS - people living with HIV/AIDS, HIV/AIDS - adult prevalence rate, and Life expectancy at birth:

In New Zealand, the Obesity - adult prevalence rate is 30.8%, Hospital bed density is 2.8 beds/1,000 populations, Physician's density is 3.03 physicians/1,000 populations, Health expenditures is 9.3%, HIV/AIDS – deaths is <100, HIV/AIDS - people living with HIV/AIDS is 3,600, HIV/AIDS - adult prevalence rate is 0.1%, and Life expectancy at birth is total population: 81.4 years; male: 79.2 years and female: 83.6 years.

6.13 New Zealand Population Projections:

The declining population is expected to continue in the years to come in New Zealand, but the population will still be growing- just at a slower rate. Current projections believe that the annual growth rate will peak in 2020 at 0.94% before gradually decreasing to 0.34% in 2050. With growth rates this small, it is unlikely that New Zealand will see large changes in their numbers. The same set of predictions believe that the population of New Zealand will be 4,834,420 in 2020, 5,213,103 in 2030, 5,502,172 in 2040, and 5,711,484 by 2050. Following figure 8 shows the population projection graph in New Zealand.

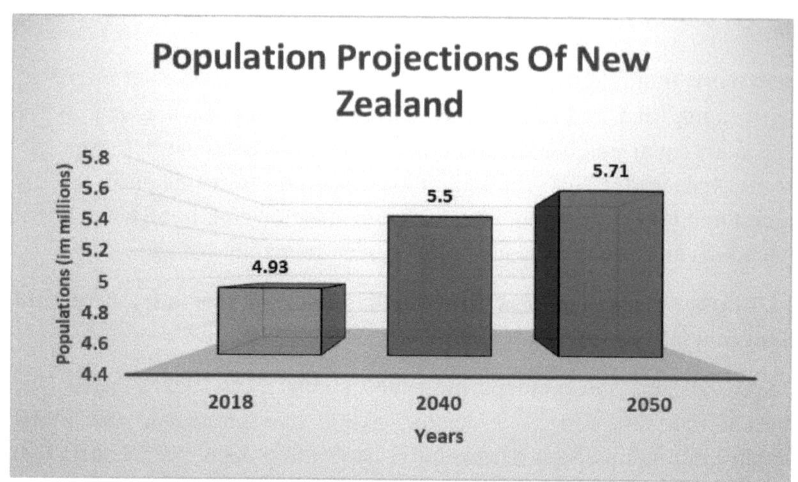

Figure 8: the population projection graph in New Zealand

6.14 School life expectancy (primary to tertiary education) and Education expenditures

Education follows the three-tier model, which includes primary schools, followed by secondary schools (high schools) and tertiary education at universities or polytechnics. The Programme for International Student Assessment ranked New Zealand's education as the seventh highest in 2009. The Education Index, published with the UN's 2014 Human Development Index and based on data from 2013, listed New Zealand at 0.917, ranked second after Australia.

School life expectancy (primary to tertiary education) is total of 19 years, (male 18 years and females 20 years) and the total educational expenditures of New Zealand's are 6.3% of GDP.

6.15 Income:

In 1982 New Zealand had the lowest per-capita income of all the developed nations surveyed by the World Bank. In 2010 the estimated gross domestic product (GDP) at purchasing power parity (PPP) per capita was roughly US$28,250, between the thirty-first and fifty-first highest for all countries. The median personal income in 2006 was $24,400. This was up from $15,600 in 1996, with the largest increases in the $50,000 to $70,000 bracket. The median income for men was $31,500, $12,400 more than women. The highest median personal income were for people identifying with the European or "other" ethnic group, while the lowest was from the Asian ethnic group. The median income for people identifying as Māori was $20,900. In 2013, the median personal income had risen slightly to $28,500.

6.16 Languages:

Various languages diversity are present (fig. 9) in New Zealand. There are such as, English (de facto official) 93.01%, Maori (de jure official) 2.09%, Samoan 0.9%, Northern Chinese 2%, Hindi 0.6%, French 0.9%, Yue 0.4%, New Zealand Sign Language (de jure official) 0.1%.

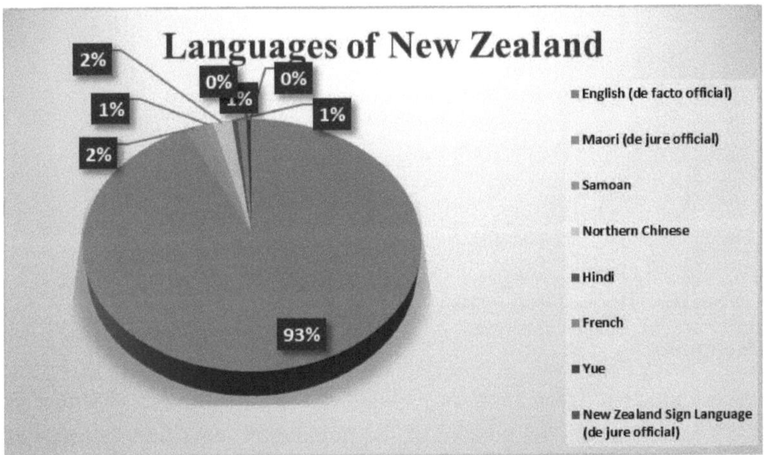

Figure 9: Various languages diversity

6.17 Ethnic groups:

New Zealand is a multi-ethnic society, and home to people of many different national origins. Originally composed solely of the Māori who arrived in the thirteenth century, the ethnic makeup of the population later became dominated by New Zealanders of European descent. In the nineteenth century, European settlers brought diseases for which the Māori had no immunity. By the 1890s, the Māori population was approximately 40 percent of its size pre-contact. The Māori population increased during the twentieth century, though it remains a minority. The 1961 New Zealand census recorded that the population was 92 percent European and 7 percent Māori, with Asian and Pacific minorities sharing the remaining 1 percent.

And the presently, that types of ethnic (European 64.1%, Maori 16.5%, Chinese 4.9%, Indian 4.7%, Samoan 3.9%, Tongan 1.8%, Cook Islands Maori 1.7%, English 0.5%, Filipino 0.9%, New Zealander 1%) are here. See the figure 10.

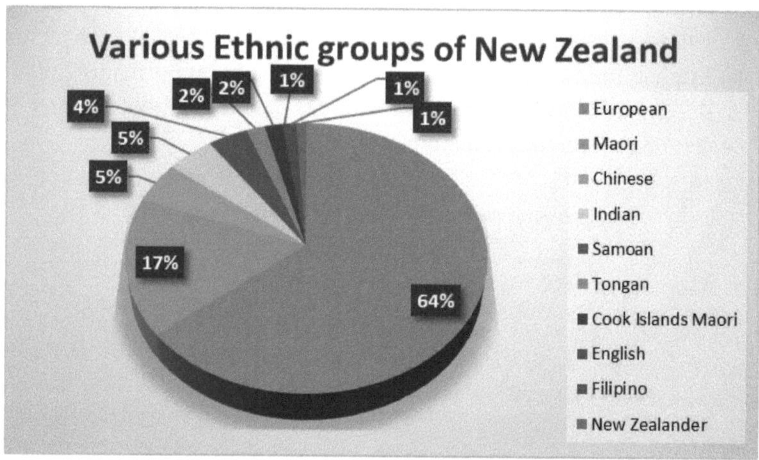

Figure 10: New Zealand is a multi-ethnic society

6.18 Religions:

The predominant religion in New Zealand is Christianity. As recorded in the 2018 census, about 38 percent of the population identified themselves as Christians, although regular church attendance is estimated at 15 percent. Another 48.5 percent indicated that they had no religion (up from 41.9 percent in 2013 and 34.7 percent in 2006) and around 7.5 percent affiliated with other religions.

The indigenous religion of the Māori population was animistic, but with the arrival of missionaries from the early nineteenth century most of the Māori population converted to Christianity. In the 2018 census, 3,699 Māori still identify themselves as adhering to "Māori religions, beliefs and philosophies". In the 2018 census, the largest reported Christian affiliations are Anglican (6.7 percent of the population), Roman Catholic (6.3 percent),

Presbyterian (4.7 percent). There are also significant numbers of Christians who identify themselves with Methodist, Pentecostal, Baptist and Latter-day Saint churches, and the New Zealand-based Rātana church has adherents among Māori.[Immigration and associated demographic change in recent decades has contributed to the growth of minority religions, especially Hinduism, Buddhism and Islam. Overviews is that the figure 11.

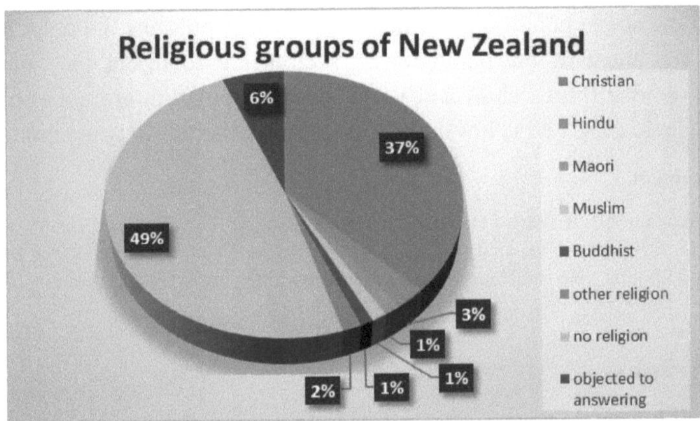

Figure 11: Religious groups of New Zealand

6.19 Unemployment rate:

The Unemployment rate of New Zealand is total: 12.7% (male: 12.4%, female: 13%). See (fig. 12) the distribution.

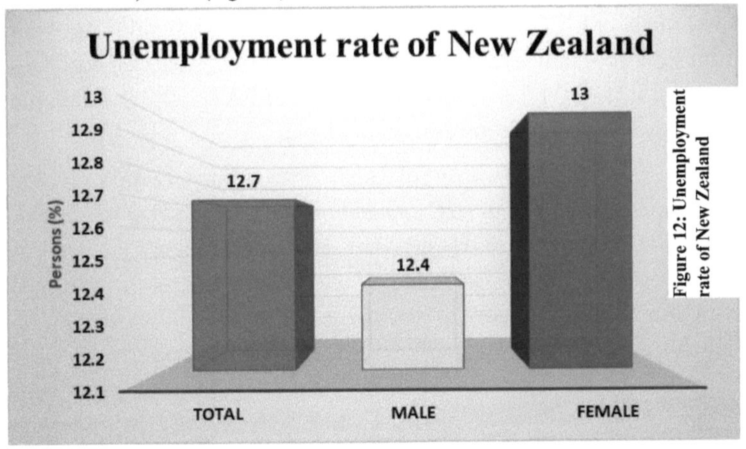

7. Conclusion:

This study based on the availability of New Zealand census data (Statistics of New Zealand), most of the demographic and some selected social demographic indicators are covered. From this discussion, the country of New Zealand is a rich country and living way is good and GDP is $199 billion and per capita income is $40,266, and the HDI value is very high 0.921. And the social conditions are not so bad. Here, presents the unity in diversity. And most of the demographic sectors are represent the good results but the one things child and youth stages population are not good conditions (increasing rate is low), it will be affected the future generation.

Acknowledgement

The author thanks to Dr. Bhaswati Das, Dr. Nandita Saikia, Dr. D.N.Das, and Dr. Milap Punia, (CSRD, JNU), Dr. Nieuwaal (Director Delta Alliance, The Netherlands) for the theoretical and applied help (teaching and valuable knowledge share) of this works.

8. Bibliography:

1) Alho, J., and Spencer, B., (2005). Statistical Demography and Forecasting. Springer-Verlag New York, ISBN: 978-0-387-28392-0.

2) Bongaarts, J., and Blanc, A.K., (2015). Estimating the current mean age of mothers at the birth of their first child from household surveys. Bongaarts and Blanc Population Health Metrics, 13 (25), 1 - 6.

3) Braverman, M. T, & Engle, M. (2009). Theory and rigor in Extension program evaluation planning. *Journal of Extension* [On-line], 47(3) Article 3FEA1. Available at: https://www.joe.org/joe/2009june/a1.php

4) Centers for Disease Control and Prevention (2009). *Quality assurance standards for HIV counseling, testing, and referral data.* Atlanta: Department of Health and Human Services, Centers of Disease Control and Prevention. Retrieved from: http://www.cdc.gov/hiv/testing/resources/guidelines/quas/over view.htm#link4

5) Dunifon, R., Duttweiler, M., Pillemer, K., Tobias, D., & Trochim, W. M. (2004). Evidence-based Extension. *Journal of Extension* [On-line], 42(2) Article 2FEA2. Available at: https://www.joe.org/joe/2004april/a2.php

6) Guba, E. G. (1981). Criteria for assessing the trustworthiness of naturalistic inquiries. *Educational Communication and Technology Journal*, 29 (2), 75-91.

7) Guba, E. G., & Lincoln, Y. S. (1981). *Effective evaluation: Improving the usefulness of evaluation results through responsive and naturalistic approaches*. San Francisco, CA: Jossey-Bass.

8) Hayward, M.D., (2020). Demography. Population Association of America (Springer), 57(4). ISSN: 1533-7790.

9) Malakar, K.D., (2020). A Basic Outline of Population Studies. Lambert, Germany. ISBN: 978-620-0-54047-8.

10) Mester, L.J., (2017). Demographics and Their Implications for the Economy and Policy. Cato Institute's 35th Annual Monetary Conference: The Future of Monetary Policy, Washington, DC, pp. 1-16.

11) Miller, L. E., & Smith, K. L. (1983). Handling nonresponse issues. *Journal of Extension* [On-line], 21(5), Available at: https://www.joe.org/joe/1983september/83-5-a7.pdf

12) Organization for Economic Co-operation and Development (2003). *Quality framework and guidelines for OECD Statistical activities, version 2003/1*. Retrieved from: http://www.oeced.org/dataoecd/26/42/21688835.pdf

13) Petit, V., (2018). Population Studies and Development from Theory to Fieldwork. Springer International Publishing, ISBN: 978-3-319-61774-9.

14) Poston, D.L., (2019). Handbook of Population. Springer International Publishing, ISBN: 978-3-030-10910-3.

15) Preston, S.H., Heuveline, P., and Guillot, M., (2001). Demography (Measuring and Modeling Population Processes). Blackwell Publishing, ISBN 1557864519.

16)   Radhakrishna, R. B., & Doamekpor, P. (2008). Strategies for generalizing findings in survey research. *Journal of Extension* [On-line] 46(2), Article 2TOT1. Available at: https://www.joe.org/joe/2008april/tt1.php

17)   Radhakrishna, R. B., & Relado, R. Z. (2009). A framework to link evaluation questions to program outcomes. *Journal of Extension* [On-line], 47(3) Article 3TOT2. Available at: https://www.joe.org/joe/2009june/tt2.php

18)   Vale, S. (2010). Statistical data quality in the UNECE, 2010 version. Statistical Division, United Nations. Retrieved from: http://unstats.un.org/unsd/dnss/docs-nquaf/UNECE-quality%20Improvement%20Programme%202010.pdf

****

Chapter – 7
Social Demography of New Zealand

Please see the chapter no 6.

****

Chapter – 8

Demographics of Auckland and the Cook Islands

Demographics of Auckland

Auckland is New Zealand's most populous city. In the 2013 census, 1,415,550 persons declared themselves as residents of the Auckland region an increase of 110,000 people or 8% since the 2006 census. Auckland accounts for about one-third (33.4%) of New Zealand's total population. Auckland is known for having a large multicultural mix, with the largest Polynesian population in the world.

While having strong natural population growth, Auckland also has significant external (from overseas) immigration partially offset by internal (within New Zealand) emigration. During the last decade up to 2011, approximately 50 people per day moved to Auckland, requiring an average of 21 new homes, and occupying in excess of one extra hectare of land a day.

Ethnicity

The proportion of Asians and other non-European immigrants has increased during the last decades due to immigration, and the removal of restrictions directly or indirectly based on ethnicity. Immigration to New Zealand is heavily concentrated towards Auckland (partly for job market reasons). This strong focus on Auckland has led the immigration services to award extra points towards immigration visa requirements for people intending to move to other parts of New Zealand.

The following table shows the ethnic profile of Auckland's population, as recorded in the 2001, 2006, and 2013 New Zealand censuses. The substantial percentage drop of 'Europeans' in 2006 was mainly caused by the increasing numbers of people from this group choosing to define themselves as 'New Zealanders', as a result of a media campaign that encouraged people to give the response 'New Zealander' even though this was not one of the groups listed on the census form. In the 2013 census fewer Europeans identify themselves as 'New Zealander', leading to a significant increase of numbers in 'Europeans'

Ethnicity	2001 census		2006 census		2013 census		2018 census	
	Number	%	Number	%	Number	%	Number	%
European	755,967	68.5	700,158	56.5	789,306	59.3	841,386	53.5
New Zealand European	685,965	62.2	611,898	49.4	696,963	52.3		
South African	9,183	0.8	12,888	1.0	15,741	1.2		
English	13,398	1.2	16,008	1.3	13,497	1.0		
British	6,468	0.6	9,603	0.8	12,087	0.9		
European (not further defined)	10,197	0.9	9,162	0.7	10,668	0.8		
Dutch	7,728	0.7	7,785	0.6	7,998	0.6		
Australian	7,200	0.7	8,637	0.7	7,065	0.5		
Asian	151,644	13.8	234,279	18.9	307,233	23.1	442,674	28.2

Chinese	65,952	6.0	92,889	7.5	112,290	8.4		
Indian	40,326	3.7	69,315	5.6	97,875	7.5		
Korean	13,272	1.2	21,354	1.7	21,981	1.7		
Filipino	6,312	0.6	9,822	0.8	20,499	1.5		
Fijian Indian	1,281	0.1	4,170	0.3	8,067	0.6		
Japanese	4,221	0.4	5,289	0.4	6,720	0.5		
Pacific peoples	154,683	14.0	177,948	14.4	194,958	14.6	243,966	15.5
Samoan	76,584	6.9	87,840	7.1	95,916	7.2		
Tongan	32,538	3.0	40,140	3.2	46,971	3.5		
Cook Islands Māori	31,077	2.8	34,374	2.8	36,549	2.7		
Niuean	16,035	1.5	17,667	1.4	18,555	1.4		

	Num	%	Num	%	Num	%	Num	%
Fijian	4,155	0.4	5,850	0.5	8,493	0.6		
Māori	127,704	11.6	137,304	11.1	142,767	10.7	181,194	11.5
Middle Eastern/Latin American/African	13,335	1.2	18,558	1.5	24,945	1.9	35,838	2.3
Other	276	<0.1	100,110	8.1	15,639	1.1	16,746	1.1
New Zealander	N/A		99,474	8.0	14,904	1.1		
Total people stated	1,102,818		1,239,054		1,331,427		1,517,718	
Not elsewhere included	57,453	5.0	65,907	5.1	84,123	5.9	0	0.0

Ethnic groups by Auckland local board area, 2018 census

Local board area	European		Maori		Pacific		Asian		MELAA		Other	
	Num.	%	Num.	%	Num.	%	Num.	%	Num.	%	Num.	%
Rodney	59,013	88.9	7,551	11.4	2,340	3.5	3,756	5.7	450	0.7	822	1.2

Hibiscus and Bays	84,057	80.8	6,735	6.5	2,205	2.1	16,626	16.0	1,551	1.5	1,206	1.2
Upper Harbour	34,746	55.3	3,210	5.1	1,530	2.4	24,867	39.6	1,887	3.0	897	1.4
Kaipatiki	51,633	58.5	7,680	8.7	5,379	6.1	29,034	32.9	2,718	3.1	1,149	1.3
Devonport-Takapuna	40,152	69.3	3,192	5.5	1,443	2.5	15,249	26.3	1,476	2.5	681	1.2
Henderson-Massey	57,633	48.7	20,319	17.2	24,771	20.9	32,523	27.5	3,087	2.6	1,416	1.2
Waitākere Ranges	38,823	74.5	6,621	12.7	6,093	11.7	7,275	14.0	936	1.8	642	1.2
Great Barrier	855	91.3	192	20.5	24	2.6	15	1.6	0	0.0	18	1.9
Waiheke	8,055	88.9	1,035	11.4	342	3.8	366	4.0	315	3.5	108	1.2
Waitemata	49,950	60.3	5,034	6.1	4,053	4.9	26,103	31.5	3,912	4.7	828	1.0
Whau	32,040	40.4	7,845	9.9	14,817	18.7	31,959	40.3	2,286	2.9	780	1.0
Albert-Eden	58,899	59.7	7,005	7.1	7,653	7.8	31,524	32.0	2,748	2.8	990	1.0
Puketapapa	19,356	33.6	3,462	6.0	8,775	15.2	28,266	49.1	2,163	3.8	510	0.9

	Num.	%	Num.	%	Num.	%	Num.	%	Num.	%	Num.	%
Orakei	61,221	72.6	4,815	5.7	2,676	3.2	19,296	22.9	2,127	2.5	825	1.0
Maungakiekie-Tamaki	33,438	43.8	10,656	14.0	19,602	25.7	21,309	27.9	1,680	2.2	711	0.9
Howick	64,776	46.0	8,052	5.7	8,028	5.7	65,541	46.5	3,540	2.5	2,010	1.4
Mangere-Otahuhu	14,976	19.1	12,861	16.4	46,578	59.4	14,925	19.0	603	0.8	444	0.6
Otara-Papatoetoe	14,142	16.6	13,392	15.7	39,198	46.0	29,880	35.1	882	1.0	498	0.6
Manurewa	27,942	29.2	24,846	26.0	34,707	36.3	24,345	25.4	1,962	2.1	645	0.7
Papakura	28,305	49.1	15,438	26.8	9,750	16.9	13,497	23.4	918	1.6	588	1.0
Franklin	61,371	82.0	11,247	15.0	4,008	5.4	6,324	8.5	585	0.8	969	1.3

Birthplace by Auckland local board area, 2013 census

Local board area	New Zealand		Australia		Pacific Islands		British Isles		Europe		North America		Asia		Middle East and Africa		Latin America and Other	
	Num.	%	Num.	%	Num.	%	Num.	%	Num.	%	Num.	%	Num.	%	Num.	%	Num.	%

Rodney	39,678	77.8	936	1.8	540	1.1	5,763	11.3	1,098	2.2	549	1.1	1,272	2.5	1,092	2.1	90	0.2
Hibiscus and Bays	53,805	63.0	1,476	1.7	624	0.7	13,443	15.7	2,574	3.0	912	1.1	5,499	6.4	6,717	7.9	321	0.4
Upper Harbour	27,525	54.1	699	1.4	762	1.5	4,296	8.4	1,260	2.5	438	0.9	11,562	22.7	4,140	8.1	168	0.3
Kaipatiki	45,189	58.1	1,251	1.6	2,358	3.0	5,991	7.7	2,238	2.9	648	0.8	15,936	20.5	3,711	4.8	432	0.6
Devonport-Takapuna	32,226	60.5	1,113	2.1	522	1.0	5,919	11.1	1,728	3.2	759	1.4	8,598	16.1	2,073	3.9	315	0.6
Henderson-Massey	64,248	64.2	1,230	1.2	10,536	10.5	4,698	4.7	2,232	2.2	501	0.5	13,293	13.3	2,931	2.9	399	0.4
Waitākere Ranges	31,902	70.6	756	1.7	2,265	5.0	4,806	10.6	1,284	2.8	498	1.1	2,454	5.4	1,038	2.3	186	0.4
Great Barrier	663	81.3	21	2.6	3	0.4	72	8.8	21	2.6	18	2.2	9	1.1	9	1.1	0	0.0
Waiheke	5,442	70.3	213	2.8	69	0.9	1,107	14.3	321	4.1	195	2.5	174	2.2	147	1.9	75	1.0
Waitemata	37,689	53.1	1,539	2.2	1,644	2.3	5,190	7.3	2,763	3.9	1,323	1.9	17,589	24.8	2,190	3.1	1,092	1.5
Whau	54,834	54.5	1,584	1.2	3,153	11.4	5,379	4.4	1,899	1.5	966	0.5	18,618	23.5	2,085	2.9	465	0.4

Local board	n	%	n	%	n	%	n	%	n	%	n	%	n	%	n	%	n	%
Albert-Eden	54,834	61.6	1,584	1.8	3,153	3.5	5,379	6.0	1,899	2.1	966	1.1	18,618	20.9	2,085	2.3	465	0.5
Puketapapa	24,192	48.3	513	1.0	5,274	10.5	1,773	3.5	645	1.3	270	0.5	15,507	31.4	1,683	3.4	204	0.4
Orakei	49,224	65.4	1,815	2.4	1,230	1.6	6,621	8.8	2,679	3.6	1,083	1.4	9,633	12.8	2,526	3.4	513	0.7
Maungakiekie-Tamaki	38,955	60.8	798	1.2	8,211	12.8	2,496	3.9	1,263	2.0	429	0.7	10,164	15.9	1,449	2.3	267	0.4
Howick	60,189	49.4	1,398	1.1	4,815	4.0	8,730	7.2	2,184	1.8	723	0.6	32,928	27.0	10,488	8.6	399	0.3
Mangere-Otahuhu	36,519	56.8	588	0.9	21,858	34.0	810	1.3	237	0.4	183	0.3	3,477	5.4	501	0.8	84	0.1
Otara-Papatoetoe	36,642	53.4	567	0.8	20,577	30.0	870	1.3	246	0.4	165	0.2	8,598	12.5	837	1.2	93	0.1
Manurewa	48,492	64.6	744	1.0	13,962	18.6	2,082	2.8	429	0.6	198	0.3	6,834	9.1	2,193	2.9	93	0.1
Papakura	32,337	76.5	555	1.3	2,463	5.8	1,950	4.6	429	1.0	174	0.4	3,162	7.5	1,107	2.6	105	0.2
Franklin	48,744	79.3	981	1.6	1,083	1.8	5,502	9.0	954	1.6	378	0.6	2,085	3.4	1,623	2.6	84	0.1

Religion:

Religion of usually resident population, 2001–13 Census

Religion	2001 census		2006 census		2013 census	
	Number	%	Number	%	Number	%
Christian	604,713	56.98	636,405	52.96	615,936	47.56
Catholic	153,678	14.48	169,881	14.14	172,110	13.29
Anglican	147,993	13.94	141,522	11.78	117,843	9.10
Presbyterian, Congregational and Reformed	110,490	10.41	109,539	9.12	95,892	7.40
Christian (not further defined)	63,180	5.95	63,714	5.30	78,480	6.06
Methodist	47,034	4.43	50,442	4.20	46,770	3.61
Pentecostal	23,238	2.19	31,104	2.59	31,386	2.42
Baptist	18,546	1.75	21,495	1.79	21,237	1.64
Latter-day Saints	16,632	1.57	19,230	1.60	19,374	1.50
Other Christian religions	32,796	3.09	37,560	3.13	39,240	3.03
Hindu	25,788	2.43	45,327	3.77	61,458	4.75
Buddhist	22,722	2.14	29,217	2.43	32,778	2.53
Islam/Muslim	15,318	1.44	23,688	1.97	31,158	2.41

Sikh	3,225	0.30	6,177	0.51	11,715	0.90
Māori Christian	14,481	1.36	14,574	1.21	11,649	0.90
Ratana	12,585	1.19	12,438	1.04	10,122	0.78
Ringatu	1,977	0.19	2,253	0.19	1,617	0.12
Other Maori Christian religions	153	0.01	141	0.01	114	0.01
Spiritualism and New Age religions	4,854	0.46	5,907	0.49	5,238	0.40
Judaism/Jewish	3,132	0.30	3,315	0.28	3,099	0.24
Other religions	4,767	0.45	5,541	0.46	5,637	0.44
Total people with at least one religious affiliation	692,691	65.27	763,191	63.51	772,095	59.61
No religion	308,592	29.08	390,411	32.49	489,915	37.83
Object to answering	68,601	6.46	67,302	5.60	48,585	3.75

Future growth

Auckland is expecting substantial population growth via immigration and natural population increases (which contribute to growth at about one-third and two-thirds, respectively), and is set to grow to an estimated 2 million inhabitants by 2050 (a compounded annual growth rate of 1.2% vs the 2013 number above). This substantial increase in population will have a major impact on transport, housing and other infrastructure that is in many cases already considered under pressure. It is also feared by some organisations, such as the Auckland Regional Council, that urban sprawl will result from the growth and, as a result, that it is necessary to address this proactively in planning policy.

A 'Regional Growth Strategy' has been adopted that sees limits on further subdivision and intensification of existing use as its main sustainability measures. This policy is contentious, as it naturally limits the uses of private land, especially the subdivision of urban fringe properties, by setting 'Metropolitan Urban Limits' in planning documents like the District Plan. According to the 2006 Census projections, the medium-variant scenario shows that the population is projected to continue growing, to reach 1.93 million by 2031. The high-variant scenario shows the region's population growing to over two million by 2031.

Demographics of the Cook Islands

The following demographic statistics are from the CIA World Factbook, unless otherwise indicated.

- Population
  - 9,290
- Age structure (2017 est.)
  - 0–14 years: 21.12% (male 1,154/female 1,025)
  - 15–24 years: 16.63% (male 929/female 806)
  - 25–54 years: 38.09% (male 1,876/female 1,867)
  - 55–64 years: 11.99% (male 569/female 494)
  - 65 years and over: 12.16% (male 551/female 567)
- Population growth rate
  - -2.79%
- Birth rate
  - 14 births/1,000 population
- Death rate
  - 8.4 deaths/1,000 population
- Infant mortality rate
  - Total: 13 deaths/1,000 live births
  - Male: 15.8 deaths/1,000 live births
  - Female: 10.1 deaths/1,000 live births
- Life expectancy at birth

- o   Total population: 76 years
- o   Male: 73.2 years
- o   Female: 79 years (2017 est.)
- Total fertility rate
  - o 2.19 children born/woman
- Nationality
  - o Cook Islander(s) (Noun)
  - o Cook Islander (Adjective)
- Ethnic groups
  - o Cook Island Maori (Polynesian) 81.3%
  - o part Cook Island Maori 6.7%
  - o Other 11.9%
- Religions
  - o Protestant 62.8%
    - Cook Islands Christian Church 49.1%
    - Seventh-day Adventist 7.9%,
    - Assemblies of God 3.7%
    - Apostolic Church 2.1%),
  - o Roman Catholic 17%
  - o Mormon 4.4%,
  - o Other 8%
    - This "Other" group includes smaller Christian denominations, and mostly non-indigenous adherents of Hinduism, Buddhism, and Islam, as well as the irreligious.
  - o None 5.6%
  - o No response 2.2%
- Languages
  - o English (official) 86.4%
  - o Cook Islands Maori (Rarotongan) (official) 76.2%
  - o Other 8.3%

****

Chapter – 9

Statistics New Zealand

Statistics New Zealand, branded as Stats NZ, is the public service department of New Zealand charged with the collection of statistics related to the economy, population, and society of New Zealand. To this end, Stats NZ produces censuses and surveys.

Statistics New Zealand employs people with a variety of skills, including statisticians, mathematicians, computer science specialists, accountants, economists, demographers, sociologists, geographers, social psychologists, and marketers.

There are seven organizational subgroups each managed by a Deputy Government Statistician:

- Macro-economic and Environment Statistics studies prices, national accounts, develops macro-economic statistics, does the government and international accounts, and ANZSIC 06 implementation.
- Social and Population Statistics study population, social conditions, the standard of the living, census, and has a census planning manager as well as statisticians helping to develop new social statistics measures.
- Standards and Methods study statistical methods, statistical education and research, solutions and capabilities, information management, and develops new methodologies.
- Collections and Dissemination services clients. There is one general manager in Christchurch and one in Auckland. It develops products and services, manages publishing and customer services.
- Organization Direction maintains contacts with key government officials, do internal audits and business planning, manage international relations and the Official Statistics System (OSS), and advises on Māori affairs.
- Industry and Labour Statistics studies business indicators, finance and performance, agriculture, energy, and work knowledge and skills.

- Organisation Development focuses on services for the agency itself, including information technology management, quality assurance, application development and support, finances, corporate support, and human resources.
- Many of the agency's powers, duties, and responsibilities are governed by acts of the New Zealand Parliament. The agency is a state sector organization of New Zealand operating under the authority of the Statistics Act 1975.

Reports offered by Statistics New Zealand

Population

Census counts, Migration, Estimates & projections, Births

Industry Sectors

Retail trade, Construction, Information technology & communications, Agriculture, Manufacturing & production, Wholesale trade, Film & television, Science & biotechnology, Energy, Horticulture

Economic indicators

Prices indexes, Balance of payments, CPI (inflation), Gross Domestic Product, National Accounts

Education & training

Secondary education, Primary education, Tertiary education

Businesses

Business characteristics, Business growth & innovation, Business finance, Research and development

Work, income & spending

Employment, Income, Strikes, Consumer spending

People & communities

Households, Families, Marriages & relationships, Pacific peoples, Language, Maori, Women, Children, Divorces

Government finance

Local government, Central government, Crown research institutes, District health boards

Tourism

Accommodation, Tourism

Health

Abortion, Injuries, Disabilities, Life expectancy, Gambling

Environment

Natural resources, Manufacturing energy use, Sustainable development

Imports & exports

Overseas cargo, Exports, Imports

Corporate

Corporate

Alcohol and Tobacco Consumption

Alcohol and Tobacco Consumption

Crime

Crime & justice

Households

Household Internet Access

Source:

Statistics New Zealand website (August 2009)

****

Chapter – 10

2018 New Zealand census

The 2018 New Zealand census was the thirty-fourth national census in New Zealand, which took place on Tuesday 6 March 2018. The population of New Zealand was counted as 4,699,755 – an increase of 457,707 (10.79%) over the 2013 census.

Results from the 2018 census were released to the public on 23 September 2019, from the Statistics New Zealand website. The next New Zealand census is set to be held in March 2023.

The 2018 census collected data on the following topics:

Population structure

- Absentees
- Age
- Legally registered relationship status
- Name
- Number of children born
- Partnership status in current relationship
- Number of occupants on census night
- Sex

Location

- Dwelling address
- Census night address
- Usual residence
- Usual residence one year ago
- Years at usual residence

Culture and identity

- Birthplace
- Ethnicity
- Iwi affiliation
- Languages spoken
- Māori descent

- Religious affiliation
- Years since arrival in New Zealand

Education and training

- Field of study
- Highest qualification
- Highest secondary school qualification
- Level of post-school qualification
- Study participation

Work

- Hours worked in employment per week
- Industry
- Occupation
- Sector of ownership
- Status in employment
- Unpaid activities
- Work and labour force status
- Workplace address

Income

- Sources of personal income
- Total personal income

Families and households

- Child dependency status
- Extended families
- Family type
- Household composition

Housing

- Access to basic amenities
- Access to telecommunication systems
- Dwelling counts (occupied, unoccupied, under construction)
- Dwelling dampness indicator
- Dwelling mould indicator
- Individual home ownership

- Main types of heating
- Number of rooms
- Number of bedrooms
- Occupied dwelling type
- Sector of landlord
- Tenure of household
- Weekly rent paid by households

Transport

- Education institution address
- Main means of travel to education
- Main means of travel to work
- Number of motor vehicles

Health and disability

- Cigarette smoking behaviour
- Disability/activity limitations

Statistics New Zealand annually conducts population projections for New Zealand as a whole, which are based on data from the previous census (in this case, the 2013 census) and calculated using a cohort-component method. Population projections also take into consideration births, deaths, and net migration. In 2016, New Zealand's population at the time of the 2018 census was projected to be between 4,807,000 and 4,944,000.

Data uses fixed random rounding to protect confidentiality; each data point is rounded either to the nearest multiple of 3 (2/3 chance) or the next-nearest multiple of 3 (1/3 chance).

The census usually-resident population count of New Zealand is a count of all people who usually live in and were present in the country on census night (6 March 2018) and excludes overseas visitors and New Zealand residents who are temporarily overseas. Due to the high rate of non-response in the census, the published results combine answers from census forms with data from the 2013 Census and from government administrative data. Reports from an External Data Quality Review Panel include quality ratings for each variable, taking the added data into account.

Chapter – 11

Bibliography

1) Alho, J., and Spencer, B., (2005). Statistical Demography and Forecasting. Springer-Verlag New York, ISBN: 978-0-387-28392-0.

2) Ben J. Wattenberg (2004), How the New Demography of Depopulation Will Shape Our Future. Chicago: R. Dee, ISBN 1-56663-606-X

3) Bongaarts, J., and Blanc, A.K., (2015). Estimating the current mean age of mothers at the birth of their first child from household surveys. Bongaarts and Blanc Population Health Metrics, 13 (25), 1 - 6.

4) Braverman, M. T, & Engle, M. (2009). Theory and rigor in Extension program evaluation planning. *Journal of Extension* [On-line], 47(3) Article 3FEA1. Available at: https://www.joe.org/joe/2009june/a1.php

5) Centers for Disease Control and Prevention (2009). *Quality assurance standards for HIV counseling, testing, and referral data.* Atlanta: Department of Health and Human Services, Centers of Disease Control and Prevention. Retrieved from: http://www.cdc.gov/hiv/testing/resources/guidelines/quas/overview.htm#link4

6) Dunifon, R., Duttweiler, M., Pillemer, K., Tobias, D., & Trochim, W. M. (2004). Evidence-based Extension. *Journal of Extension* [On-line], 42(2) Article 2FEA2. Available at: https://www.joe.org/joe/2004april/a2.php

7) Gavrilov L.A., Gavrilova N.S. 2010. Demographic Consequences of Defeating Aging. Rejuvenation Research, 13(2-3): 329–334.

8) Gavrilova N.S., Gavrilov L.A. 2011. Ageing and Longevity: Mortality Laws and Mortality Forecasts for Ageing Populations [In Czech: Stárnutí a dlouhověkost: Zákony a prognózy úmrtnosti pro stárnoucí populace]. Demografie, 53(2): 109–128.

9) Glad, John. 2008. Future Human Evolution: Eugenics in the Twenty-First Century. Hermitage Publishers, ISBN 1-55779-154-6

10)   Guba, E. G. (1981). Criteria for assessing the trustworthiness of naturalistic inquiries. *Educational Communication and Technology Journal*, 29 (2), 75-91.

11)   Guba, E. G., & Lincoln, Y. S. (1981). *Effective evaluation: Improving the usefulness of evaluation results through responsive and naturalistic approaches.* San Francisco, CA: Jossey-Bass.

12)   Hayward, M.D., (2020). Demography. Population Association of America (Springer), 57(4). ISSN: 1533-7790.

13)   Joe McFalls (2007), Population: A Lively Introduction, Population Reference Bureau

14)   Josef Ehmer, Jens Ehrhardt, Martin Kohli (Eds.): Fertility in the History of the 20th Century: Trends, Theories, Policies, Discourses. Historical Social Research 36 (2), 2011.

15)   Leonid A. Gavrilov & Natalia S. Gavrilova (1991), The Biology of Life Span: A Quantitative Approach. New York: Harwood Academic Publisher, ISBN 3-7186-4983-7

16)   Malakar, K.D., (2020). A Basic Outline of Population Studies. Lambert, Germany. ISBN: 978-620-0-54047-8.

17)   Mester, L.J., (2017). Demographics and Their Implications for the Economy and Policy. Cato Institute's 35th Annual Monetary Conference: The Future of Monetary Policy, Washington, DC, pp. 1-16.

18)   Miller, L. E., & Smith, K. L. (1983). Handling nonresponse issues. *Journal of Extension* chapterOn-line], 21(5), Available at: https://www.joe.org/joe/1983september/83-5-a7.pdf

19)   Organization for Economic Co-operation and Development (2003). *Quality framework and guidelines for OECD Statistical activities, version 2003/1.* Retrieved from: http://www.oeced.org/dataoecd/26/42/21688835.pdf

20)   Paul Demeny and Geoffrey McNicoll (Eds.). 2003. The Encyclopedia of Population. New York, Macmillan Reference USA, vol.1, 32-37

21)   Paul R. Ehrlich (1968), The Population Bomb Controversial Neo-Malthusianist pamphlet

22)   Petit, V., (2018). Population Studies and Development from Theory to Fieldwork. Springer International Publishing, ISBN: 978-3-319-61774-9.

23)   Phillip Longman (2004), The Empty Cradle: how falling birth rates threaten global prosperity and what to do about it

24)   Poston, D.L., (2019). Handbook of Population. Springer International Publishing, ISBN: 978-3-030-10910-3.

25)   Preston, S.H., Heuveline, P., and Guillot, M., (2001). Demography (Measuring and Modeling Population Processes). Blackwell Publishing, ISBN 1557864519.

26)   Preston, Samuel, Patrick Heuveline, and Michel Guillot. 2000. Demography: Measuring and Modeling Population Processes. Blackwell Publishing.

27)   Radhakrishna, R. B., & Doamekpor, P. (2008). Strategies for generalizing findings in survey research. *Journal of Extension* [On-line] 46(2), Article 2TOT1. Available at: https://www.joe.org/joe/2008april/tt1.php

28)   Radhakrishna, R. B., & Relado, R. Z. (2009). A framework to link evaluation questions to program outcomes. *Journal of Extension* [On-line], 47(3) Article 3TOT2. Available at: https://www.joe.org/joe/2009june/tt2.php

29)   Sven Kunisch, Stephan A. Boehm, Michael Boppel (eds) (2011). From Grey to Silver: Managing the Demographic Change Successfully, Springer-Verlag, Berlin Heidelberg, ISBN 978-3-642-15593-2

30)   Uhlenberg P. (Editor), (2009) International Handbook of the Demography of Aging, New York: Springer-Verlag, pp. 113–131.

31)   Vale, S. (2010). Statistical data quality in the UNECE, 2010 version. Statistical Division, United Nations. Retrieved from: http://unstats.un.org/unsd/dnss/docs-nquaf/UNECE-quality%20Improvement%20Programme%202010.pdf

The End

Printed by Books on Demand GmbH, Norderstedt / Germany